I0486498

# Thermodynamic Properties of Supercritical Steam

Supercritical Steam Tables

Ashok Malhotra

Copyright © 2006 by Ashok Malhotra.

Published by SteamCenter.com
82, Third Avenue
G Defence Colony
Jaipur 302 021

In association with

Lulu.com
3131 RDU Center, Suite 235,
Morrisville, North Carolina 27560
**www.Lulu.com**

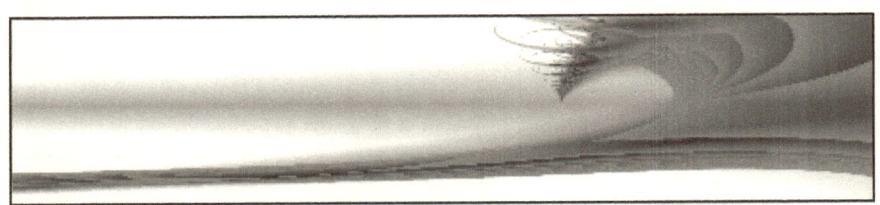

# Introduction

Supercritical is a thermodynamic expression describing the state of a fluid above a certain pressure when there exists no clear distinction between the liquid and gaseous phases. The pressure at which such a transition to the supercritical state takes place is known as the critical pressure. For water this pressure, $p_c$, is 22.064 Mpa. The corresponding saturation temperature is known as the critical temperature, $t_c$ and for water it is 647.096 K. The term critical has been used for this thermodynamic state because it is a singularity in the fluid property states. Fluids below the supercritical pressure are termed as subcritical whereas those above the critical pressure are regarded as supercritical.

As a fluid is heated in the supercritical state it undergoes a continuous transition from a liquid-like state to a vapor-like state. There is no distinct temperature such as a boiling point in the supercritical state. However, a temperature known as the pseudocritical temperature may be located below which a fluid behaves as a liquid. The pseudocritical temperature may be located precisely by observing the behavior of constant pressure specific heat of the fluid. It peaks at the pseudo critical temperature as a fluid is heated at constant pressure. Thus for example the pseudocritical temperature, $t_{sc}$ of steam at 70 Mpa is 763.7 K. At other pressures the pseudocritical temperature can be estimated to within one percent by the following equations[1]

---

[1] Pranav Kumar and Ashok Malhotra, The Pseudocritical temperature of steam, Unpublished report, Department of Mechanical Engineering, IIT Delhi, December 2000

$T* = (P*)^F$

where,

$T* = t_{sc}/t_c$

$P* = p/p_c$

and

$F = 0.1248 + 0.01424\ P* - 0.0026\ (P*)^2$

**Industrial Formulations**

In the 1960s, an Industrial Formulation for thermodynamic properties of water and steam known as IFC-67 was developed[2]. It was recognized as the standard formulation for industrial use such as in power cycle calculations. Several steam tables and software being used by industry even today are based on these formulations. However, since then better standards have become available. Compared to today's requirements the IFC-67 formulations contain weaknesses.

New formulations were developed and adopted by the International Association for the Properties of Water and Steam (IAPWS) at its meeting in Erlangen, Germany in 1997 under the name IAPWS-IF97 [3]. The present supercritical steam tables are based on these formulations. In the supercritical pressure range the formulations are applicable up to a pressure of 100 Mpa at temperatures between $0^0$C and $800^0$C. This is

---

[2] International Formulation Committee of the 6th International Conference on the properties of steam 1967, The 1967 IFC Formulation for Industrial Use, Verin Deuscher Ingenieure, Dusseldorf. 1967

[3] W. Wagner et al, The IAPWS Industrial Fomulation 1997 for the Thermodynamic Properties of Water and Steam, Transactions of the ASME, p150, vol. 122, January 2000

the range of interest for steam power calculations and primarily the range of the present steam tables. Extensions for higher temperatures are available[4] (See also[5]) but these were not in preparing the present tables.

## Supercritical Steam Power

One of the primary means of increasing efficiency of steam power plants is to increase peak pressures[6] (see also[7]). Several modern plants operate at peak pressures of more than 24 Mpa and hence function as supercritical power plants. Supercritical coal fired power plants with efficiencies of around 45% have much lower emissions than subcritical plants for a given power output. Today's state of the art in supercritical coal fired power plants permits efficiencies that exceed 45%, depending on cooling conditions. Options to increase efficiency above 50 % in ultra-supercritical power plants rely on elevated steam conditions as well as on improved process and component quality. Steam conditions up to 30 MPa/600°C/620°C are achieved using steels with 12 % chromium content. Pressures of up to 31.5 MPa/620°C/620°C have been proposed using Austenite, which is a proven, but expensive, material. Nickel-based alloys, e.g. Inconel, may permit 35 MPa/720°C/720°C, yielding efficiencies of nearly 50%.

---

[4] A. Malhotra and D.M.R. Panda, Thermodynamic Properties of superheated and supercritical steam, Applied Energy, p 387, Vol 68, 2001

[5] M. Yamazaki and T. Muto, Extrapolation of IAPWS Industrial Formulation to high temperature above 800 °C, The 14th International Conference on the Properties of Water and Steam (14th ICPWS), "Water, Steam and Aqueous Solutions for Electric Power - Advances in Science and Technology -" Kyoto, Japan, August, 2004.

[6] On the Rational selection of Reheat Pressure Ratios in Supercritical Steam Cycles, A. Malhotra and R. Satyakam, ECOS 2000, University of Twente, the Netherlands, July 5-7, 2000

[7] Influence of Climatic Parameters on Optimal Design of Supercritical Power Plants, R. Satyakam and A. Malhotra, Paper No. 2000-2987, AIAA, 35th Intersociety Energy Conversion Engineering Conference, Las Vegas, 24-28 July, 2000

## Need for Supercritical SteamTables

Although excellent software programs are available for thermodynamic properties of steam much research and design work in industry still takes place by reference to property tables. There is a need to augment currently available tables in the supercritical range. Hence the present tables were developed in a convenient to read format by the author. These have now been reproduced in this book so that they may become available to other interested persons as well through the rapid medium of the Internet or otherwise.

## Units

**The following tables in S.I. units are based on IAPWS-IF97. Temperature, t, is expressed in Kelvin, The Pressure, p, is in Mpa. Specific Volume, v, is expressed in $m^3$/kg whereas internal energy, u, and enthalpy, h,  are in kJ/kg. The Specific Entropy, s, is expressed in kJ/kg K**

## SUPERCRITICAL STEAM TABLES

| p | t | v | u | h | s |
|---|---|---|---|---|---|
| Pressure, p = 22.1 Mpa | | | | | |
| 22.1 | 275 | 0.0009894 | 7.865 | 29.73 | 0.02827 |
| 22.1 | 280 | 0.0009898 | 28.45 | 50.33 | 0.1025 |
| 22.1 | 285 | 0.0009904 | 49.04 | 70.93 | 0.1754 |
| 22.1 | 290 | 0.0009913 | 69.62 | 91.53 | 0.2471 |
| 22.1 | 295 | 0.0009925 | 90.20 | 112.1 | 0.3175 |
| 22.1 | 300 | 0.0009938 | 110.8 | 132.8 | 0.3868 |
| 22.1 | 305 | 0.0009954 | 131.4 | 153.4 | 0.4550 |
| 22.1 | 310 | 0.0009972 | 152.0 | 174.0 | 0.5221 |
| 22.1 | 315 | 0.0009991 | 172.6 | 194.6 | 0.5881 |
| 22.1 | 320 | 0.001001 | 193.2 | 215.3 | 0.6532 |
| 22.1 | 325 | 0.001003 | 213.8 | 235.9 | 0.7172 |
| 22.1 | 330 | 0.001006 | 234.4 | 256.6 | 0.7803 |
| 22.1 | 335 | 0.001008 | 255.0 | 277.3 | 0.8425 |
| 22.1 | 340 | 0.001011 | 275.7 | 298.0 | 0.9039 |

| p | t | v | u | h | s |
|---|---|---|---|---|---|
| 22.1 | 345 | 0.001014 | 296.3 | 318.7 | 0.9643 |
| 22.1 | 350 | 0.001017 | 317.0 | 339.4 | 1.024 |
| 22.1 | 355 | 0.001020 | 337.6 | 360.2 | 1.083 |
| 22.1 | 360 | 0.001023 | 358.3 | 381.0 | 1.141 |
| 22.1 | 365 | 0.001027 | 379.1 | 401.7 | 1.198 |
| 22.1 | 370 | 0.001030 | 399.8 | 422.6 | 1.255 |
| 22.1 | 375 | 0.001034 | 420.6 | 443.4 | 1.311 |
| 22.1 | 380 | 0.001038 | 441.3 | 464.3 | 1.366 |
| 22.1 | 385 | 0.001042 | 462.1 | 485.2 | 1.421 |
| 22.1 | 390 | 0.001046 | 483.0 | 506.1 | 1.475 |
| 22.1 | 395 | 0.001050 | 503.9 | 527.1 | 1.528 |
| 22.1 | 400 | 0.001055 | 524.8 | 548.1 | 1.581 |
| 22.1 | 405 | 0.001059 | 545.7 | 569.1 | 1.633 |
| 22.1 | 410 | 0.001064 | 566.7 | 590.2 | 1.685 |
| 22.1 | 415 | 0.001068 | 587.7 | 611.3 | 1.736 |
| 22.1 | 420 | 0.001073 | 608.8 | 632.5 | 1.787 |
| 22.1 | 425 | 0.001079 | 629.9 | 653.7 | 1.837 |
| 22.1 | 430 | 0.001084 | 651.1 | 675.0 | 1.887 |
| 22.1 | 435 | 0.001089 | 672.3 | 696.3 | 1.936 |
| 22.1 | 440 | 0.001095 | 693.5 | 717.7 | 1.985 |

| p | t | v | u | h | s |
|------|-----|----------|--------|--------|-------|
| 22.1 | 445 | 0.001101 | 714.9 | 739.2 | 2.034 |
| 22.1 | 450 | 0.001107 | 736.2 | 760.7 | 2.082 |
| 22.1 | 455 | 0.001113 | 757.7 | 782.3 | 2.130 |
| 22.1 | 460 | 0.001119 | 779.2 | 804.0 | 2.177 |
| 22.1 | 465 | 0.001126 | 800.8 | 825.7 | 2.224 |
| 22.1 | 470 | 0.001133 | 822.5 | 847.6 | 2.271 |
| 22.1 | 475 | 0.001140 | 844.3 | 869.5 | 2.317 |
| 22.1 | 480 | 0.001147 | 866.2 | 891.5 | 2.363 |
| 22.1 | 485 | 0.001155 | 888.1 | 913.7 | 2.409 |
| 22.1 | 490 | 0.001162 | 910.2 | 935.9 | 2.455 |
| 22.1 | 495 | 0.001171 | 932.4 | 958.3 | 2.500 |
| 22.1 | 500 | 0.001179 | 954.7 | 980.8 | 2.545 |
| 22.1 | 505 | 0.001188 | 977.2 | 1003.4 | 2.590 |
| 22.1 | 510 | 0.001197 | 999.7 | 1026.2 | 2.635 |
| 22.1 | 515 | 0.001206 | 1022.5 | 1049.1 | 2.680 |
| 22.1 | 520 | 0.001216 | 1045.4 | 1072.2 | 2.725 |
| 22.1 | 525 | 0.001226 | 1068.4 | 1095.5 | 2.769 |
| 22.1 | 530 | 0.001237 | 1091.7 | 1119.0 | 2.814 |
| 22.1 | 535 | 0.001248 | 1115.1 | 1142.7 | 2.858 |
| 22.1 | 540 | 0.001260 | 1138.8 | 1166.6 | 2.903 |

Thermodynamic Properties of Supercritical Steam

| p | t | v | u | h | s |
|------|-----|----------|--------|--------|-------|
| 22.1 | 545 | 0.001272 | 1162.7 | 1190.8 | 2.947 |
| 22.1 | 550 | 0.001285 | 1186.9 | 1215.3 | 2.992 |
| 22.1 | 555 | 0.001298 | 1211.3 | 1240.0 | 3.037 |
| 22.1 | 560 | 0.001313 | 1236.0 | 1265.1 | 3.082 |
| 22.1 | 565 | 0.001328 | 1261.1 | 1290.5 | 3.127 |
| 22.1 | 570 | 0.001344 | 1286.6 | 1316.3 | 3.172 |
| 22.1 | 575 | 0.001361 | 1312.4 | 1342.5 | 3.218 |
| 22.1 | 580 | 0.001379 | 1338.8 | 1369.2 | 3.265 |
| 22.1 | 585 | 0.001399 | 1365.6 | 1396.5 | 3.311 |
| 22.1 | 590 | 0.001420 | 1393.0 | 1424.4 | 3.359 |
| 22.1 | 595 | 0.001443 | 1421.1 | 1453.0 | 3.407 |
| 22.1 | 600 | 0.001469 | 1450.0 | 1482.4 | 3.456 |
| 22.1 | 605 | 0.001497 | 1479.8 | 1512.9 | 3.507 |
| 22.1 | 610 | 0.001528 | 1510.7 | 1544.4 | 3.559 |
| 22.1 | 615 | 0.001563 | 1542.9 | 1577.5 | 3.613 |
| 22.1 | 620 | 0.001604 | 1576.8 | 1612.3 | 3.669 |
| 22.1 | 625 | 0.001652 | 1613.0 | 1649.6 | 3.729 |
| 22.1 | 630 | 0.001712 | 1652.5 | 1690.3 | 3.794 |
| 22.1 | 635 | 0.001789 | 1697.0 | 1736.6 | 3.867 |
| 22.1 | 640 | 0.001902 | 1750.7 | 1792.8 | 3.955 |

| p | t | v | u | h | s |
|------|-----|----------|--------|--------|-------|
| 22.1 | 645 | 0.002132 | 1831.2 | 1878.3 | 4.088 |
| 22.1 | 650 | 0.005334 | 2294.6 | 2412.4 | 4.913 |
| 22.1 | 655 | 0.006306 | 2389.2 | 2528.5 | 5.091 |
| 22.1 | 660 | 0.006952 | 2447.3 | 2601.0 | 5.202 |
| 22.1 | 665 | 0.007472 | 2492.2 | 2657.3 | 5.287 |
| 22.1 | 670 | 0.007920 | 2529.6 | 2704.7 | 5.358 |
| 22.1 | 675 | 0.008318 | 2562.2 | 2746.1 | 5.419 |
| 22.1 | 680 | 0.008680 | 2591.3 | 2783.1 | 5.474 |
| 22.1 | 685 | 0.009015 | 2617.7 | 2816.9 | 5.523 |
| 22.1 | 690 | 0.009327 | 2642.1 | 2848.2 | 5.569 |
| 22.1 | 695 | 0.009621 | 2664.8 | 2877.4 | 5.611 |
| 22.1 | 700 | 0.009900 | 2686.1 | 2904.9 | 5.650 |
| 22.1 | 705 | 0.01017  | 2706.4 | 2931.1 | 5.688 |
| 22.1 | 710 | 0.01042  | 2725.7 | 2956.0 | 5.723 |
| 22.1 | 715 | 0.01067  | 2744.1 | 2979.9 | 5.756 |
| 22.1 | 720 | 0.01090  | 2761.8 | 3002.8 | 5.788 |
| 22.1 | 725 | 0.01113  | 2778.9 | 3025.0 | 5.819 |
| 22.1 | 730 | 0.01135  | 2795.5 | 3046.4 | 5.849 |
| 22.1 | 735 | 0.01157  | 2811.5 | 3067.2 | 5.877 |
| 22.1 | 740 | 0.01178  | 2827.1 | 3087.4 | 5.904 |

Thermodynamic Properties of Supercritical Steam

| p | t | v | u | h | s |
|---|---|---|---|---|---|
| 22.1 | 745 | 0.01199 | 2842.2 | 3107.1 | 5.931 |
| 22.1 | 750 | 0.01219 | 2857.0 | 3126.3 | 5.957 |
| 22.1 | 755 | 0.01238 | 2871.5 | 3145.1 | 5.982 |
| 22.1 | 760 | 0.01257 | 2885.6 | 3163.5 | 6.006 |
| 22.1 | 765 | 0.01276 | 2899.5 | 3181.6 | 6.029 |
| 22.1 | 770 | 0.01295 | 2913.1 | 3199.3 | 6.053 |
| 22.1 | 775 | 0.01313 | 2926.5 | 3216.7 | 6.075 |
| 22.1 | 780 | 0.01331 | 2939.7 | 3233.8 | 6.097 |
| 22.1 | 785 | 0.01348 | 2952.7 | 3250.7 | 6.119 |
| 22.1 | 790 | 0.01366 | 2965.5 | 3267.3 | 6.140 |
| 22.1 | 795 | 0.01383 | 2978.1 | 3283.7 | 6.161 |
| 22.1 | 800 | 0.01400 | 2990.6 | 3299.9 | 6.181 |
| 22.1 | 805 | 0.01416 | 3003.0 | 3315.9 | 6.201 |
| 22.1 | 810 | 0.01433 | 3015.2 | 3331.8 | 6.220 |
| 22.1 | 815 | 0.01449 | 3027.2 | 3347.4 | 6.240 |
| 22.1 | 820 | 0.01465 | 3039.2 | 3362.9 | 6.259 |
| 22.1 | 825 | 0.01481 | 3051.1 | 3378.3 | 6.277 |
| 22.1 | 830 | 0.01497 | 3062.8 | 3393.5 | 6.296 |
| 22.1 | 835 | 0.01512 | 3074.5 | 3408.6 | 6.314 |
| 22.1 | 840 | 0.01527 | 3086.0 | 3423.6 | 6.332 |

| p | t | v | u | h | s |
|---|---|---|---|---|---|
| 22.1 | 845 | 0.01543 | 3097.5 | 3438.5 | 6.349 |
| 22.1 | 850 | 0.01558 | 3108.9 | 3453.2 | 6.367 |
| 22.1 | 855 | 0.01573 | 3120.3 | 3467.9 | 6.384 |
| 22.1 | 860 | 0.01588 | 3131.5 | 3482.4 | 6.401 |
| 22.1 | 865 | 0.01602 | 3142.7 | 3496.9 | 6.418 |
| 22.1 | 870 | 0.01617 | 3153.9 | 3511.3 | 6.434 |
| 22.1 | 875 | 0.01632 | 3165.0 | 3525.6 | 6.451 |
| 22.1 | 880 | 0.01646 | 3176.0 | 3539.8 | 6.467 |
| 22.1 | 885 | 0.01660 | 3187.0 | 3553.9 | 6.483 |
| 22.1 | 890 | 0.01675 | 3197.9 | 3568.0 | 6.499 |
| 22.1 | 895 | 0.01689 | 3208.8 | 3582.0 | 6.514 |
| 22.1 | 900 | 0.01703 | 3219.7 | 3596.0 | 6.530 |
| 22.1 | 905 | 0.01717 | 3230.5 | 3609.9 | 6.545 |
| 22.1 | 910 | 0.01731 | 3241.2 | 3623.7 | 6.561 |
| 22.1 | 915 | 0.01744 | 3252.0 | 3637.5 | 6.576 |
| 22.1 | 920 | 0.01758 | 3262.7 | 3651.3 | 6.591 |
| 22.1 | 925 | 0.01772 | 3273.4 | 3664.9 | 6.606 |
| 22.1 | 930 | 0.01785 | 3284.0 | 3678.6 | 6.620 |
| 22.1 | 935 | 0.01799 | 3294.6 | 3692.2 | 6.635 |
| 22.1 | 940 | 0.01812 | 3305.2 | 3705.8 | 6.649 |

| p | t | v | u | h | s |
|------|------|---------|--------|--------|-------|
| 22.1 | 945  | 0.01826 | 3315.8 | 3719.3 | 6.664 |
| 22.1 | 950  | 0.01839 | 3326.4 | 3732.8 | 6.678 |
| 22.1 | 955  | 0.01852 | 3336.9 | 3746.3 | 6.692 |
| 22.1 | 960  | 0.01865 | 3347.4 | 3759.7 | 6.706 |
| 22.1 | 965  | 0.01879 | 3357.9 | 3773.1 | 6.720 |
| 22.1 | 970  | 0.01892 | 3368.4 | 3786.4 | 6.734 |
| 22.1 | 975  | 0.01905 | 3378.8 | 3799.8 | 6.748 |
| 22.1 | 980  | 0.01918 | 3389.3 | 3813.1 | 6.761 |
| 22.1 | 985  | 0.01931 | 3399.7 | 3826.4 | 6.775 |
| 22.1 | 990  | 0.01944 | 3410.1 | 3839.7 | 6.788 |
| 22.1 | 995  | 0.01956 | 3420.5 | 3852.9 | 6.802 |
| 22.1 | 1000 | 0.01969 | 3430.9 | 3866.1 | 6.815 |
| 22.1 | 1005 | 0.01982 | 3441.3 | 3879.3 | 6.828 |
| 22.1 | 1010 | 0.01995 | 3451.7 | 3892.5 | 6.841 |
| 22.1 | 1015 | 0.02007 | 3462.1 | 3905.7 | 6.854 |
| 22.1 | 1020 | 0.02020 | 3472.4 | 3918.8 | 6.867 |
| 22.1 | 1025 | 0.02032 | 3482.8 | 3931.9 | 6.880 |
| 22.1 | 1030 | 0.02045 | 3493.1 | 3945.1 | 6.893 |
| 22.1 | 1035 | 0.02058 | 3503.5 | 3958.2 | 6.905 |
| 22.1 | 1040 | 0.02070 | 3513.8 | 3971.3 | 6.918 |

| p | t | v | u | h | s |
|---|---|---|---|---|---|
| 22.1 | 1045 | 0.02082 | 3524.1 | 3984.3 | 6.930 |
| 22.1 | 1050 | 0.02095 | 3534.4 | 3997.4 | 6.943 |
| 22.1 | 1055 | 0.02107 | 3544.8 | 4010.5 | 6.955 |
| 22.1 | 1060 | 0.02120 | 3555.1 | 4023.5 | 6.968 |
| 22.1 | 1065 | 0.02132 | 3565.4 | 4036.5 | 6.980 |
| 22.1 | 1070 | 0.02144 | 3575.7 | 4049.6 | 6.992 |

Pressure, p = 22.5 Mpa

| p | t | v | u | h | s |
|---|---|---|---|---|---|
| 22.5 | 275 | 0.0009892 | 7.865 | 30.12 | 0.0283 |
| 22.5 | 280 | 0.0009896 | 28.45 | 50.71 | 0.1025 |
| 22.5 | 285 | 0.0009902 | 49.02 | 71.30 | 0.1753 |
| 22.5 | 290 | 0.0009911 | 69.60 | 91.90 | 0.2470 |
| 22.5 | 295 | 0.0009923 | 90.18 | 112.5 | 0.3174 |
| 22.5 | 300 | 0.0009937 | 110.8 | 133.1 | 0.3867 |
| 22.5 | 305 | 0.0009952 | 131.3 | 153.7 | 0.4549 |
| 22.5 | 310 | 0.0009970 | 151.9 | 174.4 | 0.5219 |
| 22.5 | 315 | 0.0009989 | 172.5 | 195.0 | 0.5880 |
| 22.5 | 320 | 0.001001 | 193.1 | 215.6 | 0.6530 |
| 22.5 | 325 | 0.001003 | 213.7 | 236.3 | 0.7170 |

Thermodynamic Properties of Supercritical Steam

| p | t | v | u | h | s |
|---|---|---|---|---|---|
| 22.5 | 330 | 0.001006 | 234.3 | 257.0 | 0.7801 |
| 22.5 | 335 | 0.001008 | 254.9 | 277.6 | 0.8423 |
| 22.5 | 340 | 0.001011 | 275.6 | 298.3 | 0.9036 |
| 22.5 | 345 | 0.001014 | 296.2 | 319.0 | 0.9641 |
| 22.5 | 350 | 0.001017 | 316.9 | 339.8 | 1.024 |
| 22.5 | 355 | 0.001020 | 337.6 | 360.5 | 1.083 |
| 22.5 | 360 | 0.001023 | 358.2 | 381.3 | 1.141 |
| 22.5 | 365 | 0.001027 | 379.0 | 402.1 | 1.198 |
| 22.5 | 370 | 0.001030 | 399.7 | 422.9 | 1.255 |
| 22.5 | 375 | 0.001034 | 420.4 | 443.7 | 1.311 |
| 22.5 | 380 | 0.001038 | 441.2 | 464.6 | 1.366 |
| 22.5 | 385 | 0.001042 | 462.0 | 485.5 | 1.421 |
| 22.5 | 390 | 0.001046 | 482.9 | 506.4 | 1.475 |
| 22.5 | 395 | 0.001050 | 503.7 | 527.4 | 1.528 |
| 22.5 | 400 | 0.001054 | 524.6 | 548.4 | 1.581 |
| 22.5 | 405 | 0.001059 | 545.6 | 569.4 | 1.633 |
| 22.5 | 410 | 0.001063 | 566.5 | 590.5 | 1.685 |
| 22.5 | 415 | 0.001068 | 587.6 | 611.6 | 1.736 |
| 22.5 | 420 | 0.001073 | 608.6 | 632.8 | 1.787 |
| 22.5 | 425 | 0.001078 | 629.7 | 654.0 | 1.837 |

| p | t | v | u | h | s |
|---|---|---|---|---|---|
| 22.5 | 430 | 0.001084 | 650.9 | 675.3 | 1.887 |
| 22.5 | 435 | 0.001089 | 672.1 | 696.6 | 1.936 |
| 22.5 | 440 | 0.001095 | 693.3 | 718.0 | 1.985 |
| 22.5 | 445 | 0.001100 | 714.7 | 739.4 | 2.033 |
| 22.5 | 450 | 0.001106 | 736.0 | 760.9 | 2.081 |
| 22.5 | 455 | 0.001113 | 757.5 | 782.5 | 2.129 |
| 22.5 | 460 | 0.001119 | 779.0 | 804.2 | 2.176 |
| 22.5 | 465 | 0.001126 | 800.6 | 825.9 | 2.223 |
| 22.5 | 470 | 0.001132 | 822.3 | 847.8 | 2.270 |
| 22.5 | 475 | 0.001139 | 844.0 | 869.7 | 2.317 |
| 22.5 | 480 | 0.001147 | 865.9 | 891.7 | 2.363 |
| 22.5 | 485 | 0.001154 | 887.9 | 913.8 | 2.409 |
| 22.5 | 490 | 0.001162 | 909.9 | 936.1 | 2.454 |
| 22.5 | 495 | 0.001170 | 932.1 | 958.4 | 2.500 |
| 22.5 | 500 | 0.001178 | 954.4 | 980.9 | 2.545 |
| 22.5 | 505 | 0.001187 | 976.8 | 1003.5 | 2.590 |
| 22.5 | 510 | 0.001196 | 999.4 | 1026.3 | 2.635 |
| 22.5 | 515 | 0.001206 | 1022.1 | 1049.2 | 2.679 |
| 22.5 | 520 | 0.001215 | 1045.0 | 1072.3 | 2.724 |
| 22.5 | 525 | 0.001226 | 1068.0 | 1095.6 | 2.768 |

Thermodynamic Properties of Supercritical Steam

| p | t | v | u | h | s |
|------|-----|----------|--------|--------|-------|
| 22.5 | 530 | 0.001236 | 1091.2 | 1119.0 | 2.813 |
| 22.5 | 535 | 0.001247 | 1114.6 | 1142.7 | 2.857 |
| 22.5 | 540 | 0.001259 | 1138.3 | 1166.6 | 2.902 |
| 22.5 | 545 | 0.001271 | 1162.2 | 1190.8 | 2.946 |
| 22.5 | 550 | 0.001284 | 1186.3 | 1215.2 | 2.991 |
| 22.5 | 555 | 0.001297 | 1210.7 | 1239.9 | 3.036 |
| 22.5 | 560 | 0.001312 | 1235.4 | 1264.9 | 3.081 |
| 22.5 | 565 | 0.001327 | 1260.4 | 1290.3 | 3.126 |
| 22.5 | 570 | 0.001343 | 1285.9 | 1316.1 | 3.171 |
| 22.5 | 575 | 0.001360 | 1311.7 | 1342.3 | 3.217 |
| 22.5 | 580 | 0.001378 | 1337.9 | 1368.9 | 3.263 |
| 22.5 | 585 | 0.001397 | 1364.7 | 1396.1 | 3.310 |
| 22.5 | 590 | 0.001419 | 1392.0 | 1423.9 | 3.357 |
| 22.5 | 595 | 0.001441 | 1420.0 | 1452.4 | 3.405 |
| 22.5 | 600 | 0.001466 | 1448.8 | 1481.8 | 3.454 |
| 22.5 | 605 | 0.001494 | 1478.4 | 1512.0 | 3.505 |
| 22.5 | 610 | 0.001525 | 1509.1 | 1543.4 | 3.556 |
| 22.5 | 615 | 0.001560 | 1541.1 | 1576.2 | 3.610 |
| 22.5 | 620 | 0.001600 | 1574.8 | 1610.8 | 3.666 |
| 22.5 | 625 | 0.001647 | 1610.6 | 1647.6 | 3.725 |

| p | t | v | u | h | s |
|------|------|----------|--------|--------|-------|
| 22.5 | 630 | 0.001704 | 1649.4 | 1687.7 | 3.789 |
| 22.5 | 635 | 0.001777 | 1692.8 | 1732.8 | 3.860 |
| 22.5 | 640 | 0.001882 | 1744.2 | 1786.5 | 3.944 |
| 22.5 | 645 | 0.002069 | 1815.0 | 1861.6 | 4.061 |
| 22.5 | 650 | 0.004558 | 2218.7 | 2321.3 | 4.770 |
| 22.5 | 655 | 0.005866 | 2357.1 | 2489.1 | 5.027 |
| 22.5 | 660 | 0.006581 | 2424.4 | 2572.5 | 5.154 |
| 22.5 | 665 | 0.007132 | 2473.7 | 2634.1 | 5.247 |
| 22.5 | 670 | 0.007598 | 2513.9 | 2684.9 | 5.323 |
| 22.5 | 675 | 0.008008 | 2548.4 | 2728.6 | 5.388 |
| 22.5 | 680 | 0.008378 | 2579.0 | 2767.5 | 5.446 |
| 22.5 | 685 | 0.008718 | 2606.5 | 2802.7 | 5.497 |
| 22.5 | 690 | 0.009034 | 2631.8 | 2835.1 | 5.544 |
| 22.5 | 695 | 0.009331 | 2655.3 | 2865.2 | 5.588 |
| 22.5 | 700 | 0.009612 | 2677.3 | 2893.5 | 5.629 |
| 22.5 | 705 | 0.00988  | 2698.1 | 2920.4 | 5.667 |
| 22.5 | 710 | 0.01014  | 2717.8 | 2945.9 | 5.703 |
| 22.5 | 715 | 0.01038  | 2736.7 | 2970.3 | 5.737 |
| 22.5 | 720 | 0.01062  | 2754.8 | 2993.7 | 5.770 |
| 22.5 | 725 | 0.01085  | 2772.2 | 3016.3 | 5.801 |

Thermodynamic Properties of Supercritical Steam

| p | t | v | u | h | s |
|---|---|---|---|---|---|
| 22.5 | 730 | 0.01107 | 2789.1 | 3038.1 | 5.831 |
| 22.5 | 735 | 0.01128 | 2805.4 | 3059.3 | 5.860 |
| 22.5 | 740 | 0.01149 | 2821.2 | 3079.8 | 5.888 |
| 22.5 | 745 | 0.01170 | 2836.6 | 3099.8 | 5.915 |
| 22.5 | 750 | 0.01190 | 2851.6 | 3119.3 | 5.941 |
| 22.5 | 755 | 0.01209 | 2866.2 | 3138.3 | 5.966 |
| 22.5 | 760 | 0.01228 | 2880.6 | 3157.0 | 5.991 |
| 22.5 | 765 | 0.01247 | 2894.6 | 3175.2 | 6.015 |
| 22.5 | 770 | 0.01265 | 2908.4 | 3193.2 | 6.038 |
| 22.5 | 775 | 0.01283 | 2922.0 | 3210.8 | 6.061 |
| 22.5 | 780 | 0.01301 | 2935.3 | 3228.1 | 6.083 |
| 22.5 | 785 | 0.01319 | 2948.4 | 3245.1 | 6.105 |
| 22.5 | 790 | 0.01336 | 2961.3 | 3261.9 | 6.126 |
| 22.5 | 795 | 0.01353 | 2974.1 | 3278.5 | 6.147 |
| 22.5 | 800 | 0.01369 | 2986.7 | 3294.8 | 6.168 |
| 22.5 | 805 | 0.01386 | 2999.1 | 3311.0 | 6.188 |
| 22.5 | 810 | 0.01402 | 3011.4 | 3326.9 | 6.207 |
| 22.5 | 815 | 0.01418 | 3023.6 | 3342.7 | 6.227 |
| 22.5 | 820 | 0.01434 | 3035.6 | 3358.3 | 6.246 |
| 22.5 | 825 | 0.01450 | 3047.6 | 3373.8 | 6.265 |

| p | t | v | u | h | s |
|------|------|---------|--------|--------|-------|
| 22.5 | 830 | 0.01465 | 3059.4 | 3389.1 | 6.283 |
| 22.5 | 835 | 0.01481 | 3071.1 | 3404.3 | 6.302 |
| 22.5 | 840 | 0.01496 | 3082.8 | 3419.4 | 6.320 |
| 22.5 | 845 | 0.01511 | 3094.3 | 3434.4 | 6.337 |
| 22.5 | 850 | 0.01526 | 3105.8 | 3449.2 | 6.355 |
| 22.5 | 855 | 0.01541 | 3117.2 | 3463.9 | 6.372 |
| 22.5 | 860 | 0.01556 | 3128.5 | 3478.6 | 6.389 |
| 22.5 | 865 | 0.01570 | 3139.8 | 3493.1 | 6.406 |
| 22.5 | 870 | 0.01585 | 3151.0 | 3507.6 | 6.423 |
| 22.5 | 875 | 0.01599 | 3162.2 | 3521.9 | 6.439 |
| 22.5 | 880 | 0.01613 | 3173.2 | 3536.2 | 6.455 |
| 22.5 | 885 | 0.01627 | 3184.3 | 3550.5 | 6.472 |
| 22.5 | 890 | 0.01642 | 3195.3 | 3564.6 | 6.487 |
| 22.5 | 895 | 0.01655 | 3206.2 | 3578.7 | 6.503 |
| 22.5 | 900 | 0.01669 | 3217.1 | 3592.7 | 6.519 |
| 22.5 | 905 | 0.01683 | 3228.0 | 3606.7 | 6.534 |
| 22.5 | 910 | 0.01697 | 3238.8 | 3620.6 | 6.550 |
| 22.5 | 915 | 0.01710 | 3249.6 | 3634.4 | 6.565 |
| 22.5 | 920 | 0.01724 | 3260.3 | 3648.2 | 6.580 |
| 22.5 | 925 | 0.01737 | 3271.0 | 3662.0 | 6.595 |

Thermodynamic Properties of Supercritical Steam

| p | t | v | u | h | s |
|------|------|---------|--------|--------|-------|
| 22.5 | 930 | 0.01751 | 3281.7 | 3675.7 | 6.610 |
| 22.5 | 935 | 0.01764 | 3292.4 | 3689.3 | 6.624 |
| 22.5 | 940 | 0.01777 | 3303.0 | 3702.9 | 6.639 |
| 22.5 | 945 | 0.01791 | 3313.6 | 3716.5 | 6.653 |
| 22.5 | 950 | 0.01804 | 3324.2 | 3730.1 | 6.667 |
| 22.5 | 955 | 0.01817 | 3334.8 | 3743.6 | 6.682 |
| 22.5 | 960 | 0.01830 | 3345.3 | 3757.0 | 6.696 |
| 22.5 | 965 | 0.01843 | 3355.8 | 3770.5 | 6.710 |
| 22.5 | 970 | 0.01856 | 3366.3 | 3783.9 | 6.724 |
| 22.5 | 975 | 0.01869 | 3376.8 | 3797.3 | 6.737 |
| 22.5 | 980 | 0.01881 | 3387.3 | 3810.6 | 6.751 |
| 22.5 | 985 | 0.01894 | 3397.8 | 3823.9 | 6.764 |
| 22.5 | 990 | 0.01907 | 3408.2 | 3837.2 | 6.778 |
| 22.5 | 995 | 0.01919 | 3418.6 | 3850.5 | 6.791 |
| 22.5 | 1000 | 0.01932 | 3429.1 | 3863.8 | 6.805 |
| 22.5 | 1005 | 0.01945 | 3439.5 | 3877.0 | 6.818 |
| 22.5 | 1010 | 0.01957 | 3449.9 | 3890.2 | 6.831 |
| 22.5 | 1015 | 0.01970 | 3460.3 | 3903.4 | 6.844 |
| 22.5 | 1020 | 0.01982 | 3470.6 | 3916.6 | 6.857 |
| 22.5 | 1025 | 0.01994 | 3481.0 | 3929.8 | 6.870 |

| p | t | v | u | h | s |
|---|---|---|---|---|---|
| 22.5 | 1030 | 0.02007 | 3491.4 | 3942.9 | 6.883 |
| 22.5 | 1035 | 0.02019 | 3501.7 | 3956.1 | 6.895 |
| 22.5 | 1040 | 0.02031 | 3512.1 | 3969.2 | 6.908 |
| 22.5 | 1045 | 0.02044 | 3522.5 | 3982.3 | 6.921 |
| 22.5 | 1050 | 0.02056 | 3532.8 | 3995.4 | 6.933 |
| 22.5 | 1055 | 0.02068 | 3543.1 | 4008.5 | 6.945 |
| 22.5 | 1060 | 0.02080 | 3553.5 | 4021.5 | 6.958 |
| 22.5 | 1065 | 0.02092 | 3563.8 | 4034.6 | 6.970 |
| 22.5 | 1070 | 0.02104 | 3574.2 | 4047.7 | 6.982 |

Pressure, p = 23 Mpa

| 23 | 275 | 0.0009890 | 7.865 | 30.61 | 0.0282 |
|---|---|---|---|---|---|
| 23 | 280 | 0.0009893 | 28.44 | 51.19 | 0.1024 |
| 23 | 285 | 0.0009900 | 49.01 | 71.78 | 0.1753 |
| 23 | 290 | 0.0009909 | 69.57 | 92.37 | 0.2469 |
| 23 | 295 | 0.0009921 | 90.14 | 113.0 | 0.3173 |
| 23 | 300 | 0.0009935 | 110.7 | 133.6 | 0.3866 |
| 23 | 305 | 0.0009950 | 131.3 | 154.2 | 0.4547 |
| 23 | 310 | 0.0009968 | 151.9 | 174.8 | 0.5218 |

Thermodynamic Properties of Supercritical Steam

| p | t | v | u | h | s |
|---|---|---|---|---|---|
| 23 | 315 | 0.0009987 | 172.5 | 195.4 | 0.5878 |
| 23 | 320 | 0.001001 | 193.0 | 216.1 | 0.6528 |
| 23 | 325 | 0.001003 | 213.6 | 236.7 | 0.7168 |
| 23 | 330 | 0.001005 | 234.2 | 257.4 | 0.7799 |
| 23 | 335 | 0.001008 | 254.9 | 278.0 | 0.8421 |
| 23 | 340 | 0.001011 | 275.5 | 298.7 | 0.9034 |
| 23 | 345 | 0.001014 | 296.1 | 319.4 | 0.9638 |
| 23 | 350 | 0.001017 | 316.8 | 340.2 | 1.023 |
| 23 | 355 | 0.001020 | 337.4 | 360.9 | 1.082 |
| 23 | 360 | 0.001023 | 358.1 | 381.7 | 1.140 |
| 23 | 365 | 0.001026 | 378.8 | 402.4 | 1.198 |
| 23 | 370 | 0.001030 | 399.6 | 423.2 | 1.254 |
| 23 | 375 | 0.001034 | 420.3 | 444.1 | 1.310 |
| 23 | 380 | 0.001037 | 441.1 | 464.9 | 1.365 |
| 23 | 385 | 0.001041 | 461.9 | 485.8 | 1.420 |
| 23 | 390 | 0.001045 | 482.7 | 506.8 | 1.474 |
| 23 | 395 | 0.001050 | 503.6 | 527.7 | 1.527 |
| 23 | 400 | 0.001054 | 524.5 | 548.7 | 1.580 |
| 23 | 405 | 0.001059 | 545.4 | 569.7 | 1.633 |
| 23 | 410 | 0.001063 | 566.4 | 590.8 | 1.684 |

| p | t | v | u | h | s |
|---|---|---|---|---|---|
| 23 | 415 | 0.001068 | 587.4 | 611.9 | 1.735 |
| 23 | 420 | 0.001073 | 608.4 | 633.1 | 1.786 |
| 23 | 425 | 0.001078 | 629.5 | 654.3 | 1.836 |
| 23 | 430 | 0.001083 | 650.6 | 675.6 | 1.886 |
| 23 | 435 | 0.001089 | 671.8 | 696.9 | 1.935 |
| 23 | 440 | 0.001094 | 693.1 | 718.3 | 1.984 |
| 23 | 445 | 0.001100 | 714.4 | 739.7 | 2.033 |
| 23 | 450 | 0.001106 | 735.8 | 761.2 | 2.081 |
| 23 | 455 | 0.001112 | 757.2 | 782.8 | 2.128 |
| 23 | 460 | 0.001119 | 778.7 | 804.4 | 2.176 |
| 23 | 465 | 0.001125 | 800.3 | 826.2 | 2.223 |
| 23 | 470 | 0.001132 | 822.0 | 848.0 | 2.269 |
| 23 | 475 | 0.001139 | 843.7 | 869.9 | 2.316 |
| 23 | 480 | 0.001146 | 865.6 | 891.9 | 2.362 |
| 23 | 485 | 0.001154 | 887.5 | 914.0 | 2.408 |
| 23 | 490 | 0.001162 | 909.5 | 936.3 | 2.453 |
| 23 | 495 | 0.001170 | 931.7 | 958.6 | 2.499 |
| 23 | 500 | 0.001178 | 954.0 | 981.1 | 2.544 |
| 23 | 505 | 0.001187 | 976.4 | 1003.7 | 2.589 |
| 23 | 510 | 0.001196 | 998.9 | 1026.4 | 2.634 |

Thermodynamic Properties of Supercritical Steam

| p | t | v | u | h | s |
|---|---|---|---|---|---|
| 23 | 515 | 0.001205 | 1021.6 | 1049.3 | 2.678 |
| 23 | 520 | 0.001215 | 1044.4 | 1072.4 | 2.723 |
| 23 | 525 | 0.001225 | 1067.5 | 1095.6 | 2.767 |
| 23 | 530 | 0.001235 | 1090.7 | 1119.1 | 2.812 |
| 23 | 535 | 0.001246 | 1114.1 | 1142.7 | 2.856 |
| 23 | 540 | 0.001258 | 1137.7 | 1166.6 | 2.901 |
| 23 | 545 | 0.001270 | 1161.5 | 1190.7 | 2.945 |
| 23 | 550 | 0.001283 | 1185.6 | 1215.1 | 2.990 |
| 23 | 555 | 0.001296 | 1210.0 | 1239.8 | 3.034 |
| 23 | 560 | 0.001310 | 1234.6 | 1264.8 | 3.079 |
| 23 | 565 | 0.001325 | 1259.6 | 1290.1 | 3.124 |
| 23 | 570 | 0.001341 | 1284.9 | 1315.8 | 3.170 |
| 23 | 575 | 0.001358 | 1310.7 | 1341.9 | 3.215 |
| 23 | 580 | 0.001376 | 1336.9 | 1368.5 | 3.261 |
| 23 | 585 | 0.001396 | 1363.5 | 1395.6 | 3.308 |
| 23 | 590 | 0.001416 | 1390.8 | 1423.4 | 3.355 |
| 23 | 595 | 0.001439 | 1418.7 | 1451.8 | 3.403 |
| 23 | 600 | 0.001464 | 1447.3 | 1480.9 | 3.452 |
| 23 | 605 | 0.001491 | 1476.7 | 1511.0 | 3.502 |
| 23 | 610 | 0.001521 | 1507.2 | 1542.2 | 3.553 |

| p | t | v | u | h | s |
|---|---|---|---|---|---|
| 23 | 615 | 0.001555 | 1539.0 | 1574.8 | 3.606 |
| 23 | 620 | 0.001594 | 1572.3 | 1608.9 | 3.661 |
| 23 | 625 | 0.001640 | 1607.6 | 1645.3 | 3.720 |
| 23 | 630 | 0.001695 | 1645.7 | 1684.6 | 3.783 |
| 23 | 635 | 0.001764 | 1687.9 | 1728.5 | 3.852 |
| 23 | 640 | 0.001859 | 1736.9 | 1779.6 | 3.932 |
| 23 | 645 | 0.002015 | 1800.3 | 1846.6 | 4.036 |
| 23 | 650 | 0.002609 | 1949.6 | 2009.6 | 4.288 |
| 23 | 655 | 0.005265 | 2307.6 | 2428.7 | 4.931 |
| 23 | 660 | 0.006110 | 2392.6 | 2533.1 | 5.090 |
| 23 | 665 | 0.006710 | 2448.9 | 2603.2 | 5.196 |
| 23 | 670 | 0.007202 | 2493.1 | 2658.8 | 5.279 |
| 23 | 675 | 0.007629 | 2530.4 | 2705.9 | 5.349 |
| 23 | 680 | 0.008011 | 2563.0 | 2747.2 | 5.410 |
| 23 | 685 | 0.008359 | 2592.1 | 2784.3 | 5.464 |
| 23 | 690 | 0.008680 | 2618.6 | 2818.2 | 5.514 |
| 23 | 695 | 0.008981 | 2643.1 | 2849.6 | 5.559 |
| 23 | 700 | 0.009265 | 2665.9 | 2879.0 | 5.601 |
| 23 | 705 | 0.009534 | 2687.4 | 2906.7 | 5.641 |
| 23 | 710 | 0.009791 | 2707.8 | 2933.0 | 5.678 |

| p | t | v | u | h | s |
|---|---|---|---|---|---|
| 23 | 715 | 0.01004 | 2727.2 | 2958.1 | 5.713 |
| 23 | 720 | 0.01028 | 2745.8 | 2982.2 | 5.746 |
| 23 | 725 | 0.01050 | 2763.7 | 3005.3 | 5.779 |
| 23 | 730 | 0.01073 | 2780.9 | 3027.6 | 5.809 |
| 23 | 735 | 0.01094 | 2797.6 | 3049.2 | 5.839 |
| 23 | 740 | 0.01115 | 2813.7 | 3070.2 | 5.867 |
| 23 | 745 | 0.01135 | 2829.4 | 3090.6 | 5.895 |
| 23 | 750 | 0.01155 | 2844.7 | 3110.4 | 5.921 |
| 23 | 755 | 0.01174 | 2859.6 | 3129.8 | 5.947 |
| 23 | 760 | 0.01193 | 2874.2 | 3148.7 | 5.972 |
| 23 | 765 | 0.01212 | 2888.5 | 3167.3 | 5.996 |
| 23 | 770 | 0.01230 | 2902.5 | 3185.4 | 6.020 |
| 23 | 775 | 0.01248 | 2916.2 | 3203.3 | 6.043 |
| 23 | 780 | 0.01266 | 2929.7 | 3220.8 | 6.066 |
| 23 | 785 | 0.01283 | 2943.0 | 3238.1 | 6.088 |
| 23 | 790 | 0.01300 | 2956.1 | 3255.1 | 6.109 |
| 23 | 795 | 0.01317 | 2969.0 | 3271.8 | 6.130 |
| 23 | 800 | 0.01333 | 2981.7 | 3288.4 | 6.151 |
| 23 | 805 | 0.01350 | 2994.3 | 3304.7 | 6.171 |
| 23 | 810 | 0.01366 | 3006.7 | 3320.8 | 6.191 |

| p | t | v | u | h | s |
|----|-----|---------|--------|--------|-------|
| 23 | 815 | 0.01382 | 3019.0 | 3336.8 | 6.211 |
| 23 | 820 | 0.01397 | 3031.2 | 3352.5 | 6.230 |
| 23 | 825 | 0.01413 | 3043.2 | 3368.1 | 6.249 |
| 23 | 830 | 0.01428 | 3055.1 | 3383.6 | 6.268 |
| 23 | 835 | 0.01443 | 3067.0 | 3398.9 | 6.286 |
| 23 | 840 | 0.01458 | 3078.7 | 3414.1 | 6.304 |
| 23 | 845 | 0.01473 | 3090.4 | 3429.2 | 6.322 |
| 23 | 850 | 0.01488 | 3101.9 | 3444.2 | 6.340 |
| 23 | 855 | 0.01503 | 3113.4 | 3459.0 | 6.357 |
| 23 | 860 | 0.01517 | 3124.8 | 3473.7 | 6.375 |
| 23 | 865 | 0.01531 | 3136.2 | 3488.4 | 6.392 |
| 23 | 870 | 0.01546 | 3147.4 | 3502.9 | 6.408 |
| 23 | 875 | 0.01560 | 3158.6 | 3517.4 | 6.425 |
| 23 | 880 | 0.01574 | 3169.8 | 3531.8 | 6.441 |
| 23 | 885 | 0.01588 | 3180.9 | 3546.1 | 6.458 |
| 23 | 890 | 0.01602 | 3191.9 | 3560.3 | 6.474 |
| 23 | 895 | 0.01615 | 3202.9 | 3574.5 | 6.489 |
| 23 | 900 | 0.01629 | 3213.9 | 3588.6 | 6.505 |
| 23 | 905 | 0.01643 | 3224.8 | 3602.6 | 6.521 |
| 23 | 910 | 0.01656 | 3235.7 | 3616.6 | 6.536 |

Thermodynamic Properties of Supercritical Steam

| p | t | v | u | h | s |
|---|---|---|---|---|---|
| 23 | 915 | 0.01670 | 3246.5 | 3630.5 | 6.551 |
| 23 | 920 | 0.01683 | 3257.3 | 3644.4 | 6.566 |
| 23 | 925 | 0.01696 | 3268.1 | 3658.2 | 6.581 |
| 23 | 930 | 0.01709 | 3278.8 | 3672.0 | 6.596 |
| 23 | 935 | 0.01722 | 3289.5 | 3685.7 | 6.611 |
| 23 | 940 | 0.01736 | 3300.2 | 3699.4 | 6.626 |
| 23 | 945 | 0.01748 | 3310.9 | 3713.0 | 6.640 |
| 23 | 950 | 0.01761 | 3321.5 | 3726.6 | 6.654 |
| 23 | 955 | 0.01774 | 3332.1 | 3740.2 | 6.669 |
| 23 | 960 | 0.01787 | 3342.7 | 3753.7 | 6.683 |
| 23 | 965 | 0.01800 | 3353.3 | 3767.2 | 6.697 |
| 23 | 970 | 0.01812 | 3363.8 | 3780.7 | 6.711 |
| 23 | 975 | 0.01825 | 3374.3 | 3794.1 | 6.725 |
| 23 | 980 | 0.01838 | 3384.8 | 3807.5 | 6.738 |
| 23 | 985 | 0.01850 | 3395.3 | 3820.9 | 6.752 |
| 23 | 990 | 0.01863 | 3405.8 | 3834.2 | 6.765 |
| 23 | 995 | 0.01875 | 3416.3 | 3847.5 | 6.779 |
| 23 | 1000 | 0.01888 | 3426.7 | 3860.9 | 6.792 |
| 23 | 1005 | 0.01900 | 3437.2 | 3874.1 | 6.805 |
| 23 | 1010 | 0.01912 | 3447.6 | 3887.4 | 6.819 |

| p | t | v | u | h | s |
|---|---|---|---|---|---|
| 23 | 1015 | 0.01924 | 3458.0 | 3900.6 | 6.832 |
| 23 | 1020 | 0.01937 | 3468.4 | 3913.9 | 6.845 |
| 23 | 1025 | 0.01949 | 3478.8 | 3927.1 | 6.858 |
| 23 | 1030 | 0.01961 | 3489.2 | 3940.2 | 6.870 |
| 23 | 1035 | 0.01973 | 3499.6 | 3953.4 | 6.883 |
| 23 | 1040 | 0.01985 | 3510.0 | 3966.6 | 6.896 |
| 23 | 1045 | 0.01997 | 3520.4 | 3979.7 | 6.908 |
| 23 | 1050 | 0.02009 | 3530.7 | 3992.8 | 6.921 |
| 23 | 1055 | 0.02021 | 3541.1 | 4006.0 | 6.933 |
| 23 | 1060 | 0.02033 | 3551.5 | 4019.1 | 6.946 |
| 23 | 1065 | 0.02045 | 3561.8 | 4032.2 | 6.958 |
| 23 | 1070 | 0.02057 | 3572.2 | 4045.3 | 6.970 |

Pressure, p = 24 Mpa

| p | t | v | u | h | s |
|---|---|---|---|---|---|
| 24 | 275 | 0.0009885 | 7.865 | 31.59 | 0.0282 |
| 24 | 280 | 0.0009889 | 28.42 | 52.15 | 0.1023 |
| 24 | 285 | 0.0009896 | 48.97 | 72.72 | 0.1751 |
| 24 | 290 | 0.0009905 | 69.52 | 93.30 | 0.2467 |
| 24 | 295 | 0.0009917 | 90.08 | 113.9 | 0.3170 |

Thermodynamic Properties of Supercritical Steam

| p | t | v | u | h | s |
|---|---|---|---|---|---|
| 24 | 300 | 0.0009930 | 110.6 | 134.5 | 0.3863 |
| 24 | 305 | 0.0009946 | 131.2 | 155.1 | 0.4544 |
| 24 | 310 | 0.0009964 | 151.8 | 175.7 | 0.5214 |
| 24 | 315 | 0.0009983 | 172.3 | 196.3 | 0.5874 |
| 24 | 320 | 0.001000 | 192.9 | 216.9 | 0.6523 |
| 24 | 325 | 0.001003 | 213.5 | 237.6 | 0.7163 |
| 24 | 330 | 0.001005 | 234.1 | 258.2 | 0.7794 |
| 24 | 335 | 0.001008 | 254.7 | 278.9 | 0.8415 |
| 24 | 340 | 0.001010 | 275.3 | 299.6 | 0.9028 |
| 24 | 345 | 0.001013 | 295.9 | 320.3 | 0.9632 |
| 24 | 350 | 0.001016 | 316.6 | 341.0 | 1.023 |
| 24 | 355 | 0.001019 | 337.2 | 361.7 | 1.082 |
| 24 | 360 | 0.001023 | 357.9 | 382.4 | 1.140 |
| 24 | 365 | 0.001026 | 378.6 | 403.2 | 1.197 |
| 24 | 370 | 0.001029 | 399.3 | 424.0 | 1.254 |
| 24 | 375 | 0.001033 | 420.0 | 444.8 | 1.309 |
| 24 | 380 | 0.001037 | 440.8 | 465.7 | 1.365 |
| 24 | 385 | 0.001041 | 461.6 | 486.6 | 1.419 |
| 24 | 390 | 0.001045 | 482.4 | 507.5 | 1.473 |
| 24 | 395 | 0.001049 | 503.2 | 528.4 | 1.527 |

| p | t | v | u | h | s |
|---|---|---|---|---|---|
| 24 | 400 | 0.001053 | 524.1 | 549.4 | 1.579 |
| 24 | 405 | 0.001058 | 545.0 | 570.4 | 1.632 |
| 24 | 410 | 0.001063 | 566.0 | 591.5 | 1.683 |
| 24 | 415 | 0.001067 | 587.0 | 612.6 | 1.734 |
| 24 | 420 | 0.001072 | 608.0 | 633.7 | 1.785 |
| 24 | 425 | 0.001077 | 629.1 | 654.9 | 1.835 |
| 24 | 430 | 0.001083 | 650.2 | 676.2 | 1.885 |
| 24 | 435 | 0.001088 | 671.4 | 697.5 | 1.934 |
| 24 | 440 | 0.001094 | 692.6 | 718.9 | 1.983 |
| 24 | 445 | 0.001099 | 713.9 | 740.3 | 2.032 |
| 24 | 450 | 0.001105 | 735.2 | 761.8 | 2.080 |
| 24 | 455 | 0.001112 | 756.7 | 783.3 | 2.127 |
| 24 | 460 | 0.001118 | 778.1 | 805.0 | 2.175 |
| 24 | 465 | 0.001124 | 799.7 | 826.7 | 2.221 |
| 24 | 470 | 0.001131 | 821.3 | 848.5 | 2.268 |
| 24 | 475 | 0.001138 | 843.1 | 870.4 | 2.314 |
| 24 | 480 | 0.001145 | 864.9 | 892.4 | 2.360 |
| 24 | 485 | 0.001153 | 886.8 | 914.4 | 2.406 |
| 24 | 490 | 0.001160 | 908.8 | 936.6 | 2.452 |
| 24 | 495 | 0.001168 | 930.9 | 959.0 | 2.497 |

Thermodynamic Properties of Supercritical Steam

| p | t | v | u | h | s |
|---|---|---|---|---|---|
| 24 | 500 | 0.001177 | 953.1 | 981.4 | 2.542 |
| 24 | 505 | 0.001185 | 975.5 | 1004.0 | 2.587 |
| 24 | 510 | 0.001194 | 998.0 | 1026.7 | 2.632 |
| 24 | 515 | 0.001204 | 1020.7 | 1049.5 | 2.676 |
| 24 | 520 | 0.001213 | 1043.5 | 1072.6 | 2.721 |
| 24 | 525 | 0.001223 | 1066.4 | 1095.8 | 2.765 |
| 24 | 530 | 0.001234 | 1089.6 | 1119.2 | 2.810 |
| 24 | 535 | 0.001245 | 1112.9 | 1142.8 | 2.854 |
| 24 | 540 | 0.001256 | 1136.4 | 1166.6 | 2.898 |
| 24 | 545 | 0.001268 | 1160.2 | 1190.6 | 2.943 |
| 24 | 550 | 0.001281 | 1184.2 | 1215.0 | 2.987 |
| 24 | 555 | 0.001294 | 1208.5 | 1239.6 | 3.032 |
| 24 | 560 | 0.001308 | 1233.1 | 1264.5 | 3.076 |
| 24 | 565 | 0.001323 | 1257.9 | 1289.7 | 3.121 |
| 24 | 570 | 0.001338 | 1283.2 | 1315.3 | 3.166 |
| 24 | 575 | 0.001355 | 1308.8 | 1341.3 | 3.212 |
| 24 | 580 | 0.001373 | 1334.8 | 1367.7 | 3.257 |
| 24 | 585 | 0.001392 | 1361.3 | 1394.7 | 3.304 |
| 24 | 590 | 0.001412 | 1388.3 | 1422.2 | 3.351 |
| 24 | 595 | 0.001434 | 1416.0 | 1450.4 | 3.398 |

| p | t | v | u | h | s |
|----|-----|----------|--------|--------|-------|
| 24 | 600 | 0.001458 | 1444.3 | 1479.3 | 3.447 |
| 24 | 605 | 0.001485 | 1473.5 | 1509.1 | 3.496 |
| 24 | 610 | 0.001514 | 1503.6 | 1539.9 | 3.547 |
| 24 | 615 | 0.001547 | 1534.8 | 1571.9 | 3.599 |
| 24 | 620 | 0.001584 | 1567.5 | 1605.5 | 3.653 |
| 24 | 625 | 0.001627 | 1601.9 | 1640.9 | 3.710 |
| 24 | 630 | 0.001677 | 1638.7 | 1679.0 | 3.771 |
| 24 | 635 | 0.001740 | 1678.9 | 1720.7 | 3.837 |
| 24 | 640 | 0.001822 | 1724.3 | 1768.0 | 3.911 |
| 24 | 645 | 0.001942 | 1778.9 | 1825.5 | 4.001 |
| 24 | 650 | 0.002166 | 1856.8 | 1908.8 | 4.129 |
| 24 | 655 | 0.003550 | 2117.7 | 2202.9 | 4.579 |
| 24 | 660 | 0.005100 | 2311.3 | 2433.7 | 4.931 |
| 24 | 665 | 0.005862 | 2391.4 | 2532.1 | 5.079 |
| 24 | 670 | 0.006427 | 2447.0 | 2601.3 | 5.183 |
| 24 | 675 | 0.006896 | 2491.2 | 2656.7 | 5.265 |
| 24 | 680 | 0.007305 | 2528.7 | 2704.0 | 5.335 |
| 24 | 685 | 0.007672 | 2561.5 | 2745.6 | 5.396 |
| 24 | 690 | 0.008007 | 2590.8 | 2783.0 | 5.450 |
| 24 | 695 | 0.008317 | 2617.6 | 2817.2 | 5.500 |

| p | t | v | u | h | s |
|---|---|---|---|---|---|
| 24 | 700 | 0.008607 | 2642.3 | 2848.9 | 5.545 |
| 24 | 705 | 0.008881 | 2665.4 | 2878.6 | 5.588 |
| 24 | 710 | 0.009141 | 2687.2 | 2906.6 | 5.627 |
| 24 | 715 | 0.009390 | 2707.8 | 2933.2 | 5.664 |
| 24 | 720 | 0.009628 | 2727.5 | 2958.5 | 5.700 |
| 24 | 725 | 0.009858 | 2746.3 | 2982.8 | 5.733 |
| 24 | 730 | 0.01008 | 2764.3 | 3006.2 | 5.766 |
| 24 | 735 | 0.01029 | 2781.8 | 3028.8 | 5.796 |
| 24 | 740 | 0.01050 | 2798.6 | 3050.6 | 5.826 |
| 24 | 745 | 0.01070 | 2814.9 | 3071.8 | 5.855 |
| 24 | 750 | 0.01090 | 2830.8 | 3092.4 | 5.882 |
| 24 | 755 | 0.01109 | 2846.2 | 3112.4 | 5.909 |
| 24 | 760 | 0.01128 | 2861.3 | 3132.0 | 5.935 |
| 24 | 765 | 0.01146 | 2876.0 | 3151.1 | 5.960 |
| 24 | 770 | 0.01164 | 2890.4 | 3169.8 | 5.984 |
| 24 | 775 | 0.01182 | 2904.6 | 3188.2 | 6.008 |
| 24 | 780 | 0.01199 | 2918.4 | 3206.2 | 6.031 |
| 24 | 785 | 0.01216 | 2932.1 | 3223.9 | 6.054 |
| 24 | 790 | 0.01233 | 2945.5 | 3241.3 | 6.076 |
| 24 | 795 | 0.01249 | 2958.7 | 3258.5 | 6.097 |

| p | t | v | u | h | s |
|---|---|---|---|---|---|
| 24 | 800 | 0.01265 | 2971.7 | 3275.4 | 6.119 |
| 24 | 805 | 0.01281 | 2984.5 | 3292.0 | 6.139 |
| 24 | 810 | 0.01297 | 2997.2 | 3308.5 | 6.160 |
| 24 | 815 | 0.01313 | 3009.7 | 3324.8 | 6.180 |
| 24 | 820 | 0.01328 | 3022.1 | 3340.8 | 6.199 |
| 24 | 825 | 0.01343 | 3034.4 | 3356.8 | 6.219 |
| 24 | 830 | 0.01358 | 3046.6 | 3372.5 | 6.238 |
| 24 | 835 | 0.01373 | 3058.6 | 3388.1 | 6.256 |
| 24 | 840 | 0.01388 | 3070.5 | 3403.5 | 6.275 |
| 24 | 845 | 0.01402 | 3082.3 | 3418.8 | 6.293 |
| 24 | 850 | 0.01416 | 3094.1 | 3434.0 | 6.311 |
| 24 | 855 | 0.01431 | 3105.7 | 3449.1 | 6.329 |
| 24 | 860 | 0.01445 | 3117.3 | 3464.1 | 6.346 |
| 24 | 865 | 0.01459 | 3128.8 | 3478.9 | 6.363 |
| 24 | 870 | 0.01473 | 3140.2 | 3493.7 | 6.380 |
| 24 | 875 | 0.01486 | 3151.6 | 3508.3 | 6.397 |
| 24 | 880 | 0.01500 | 3162.9 | 3522.9 | 6.414 |
| 24 | 885 | 0.01514 | 3174.1 | 3537.4 | 6.430 |
| 24 | 890 | 0.01527 | 3185.3 | 3551.8 | 6.446 |
| 24 | 895 | 0.01540 | 3196.4 | 3566.1 | 6.462 |

Thermodynamic Properties of Supercritical Steam

| p | t | v | u | h | s |
|---|---|---|---|---|---|
| 24 | 900 | 0.01554 | 3207.5 | 3580.4 | 6.478 |
| 24 | 905 | 0.01567 | 3218.5 | 3594.6 | 6.494 |
| 24 | 910 | 0.01580 | 3229.5 | 3608.7 | 6.510 |
| 24 | 915 | 0.01593 | 3240.4 | 3622.8 | 6.525 |
| 24 | 920 | 0.01606 | 3251.3 | 3636.8 | 6.540 |
| 24 | 925 | 0.01619 | 3262.2 | 3650.7 | 6.555 |
| 24 | 930 | 0.01632 | 3273.0 | 3664.6 | 6.570 |
| 24 | 935 | 0.01644 | 3283.8 | 3678.5 | 6.585 |
| 24 | 940 | 0.01657 | 3294.6 | 3692.3 | 6.600 |
| 24 | 945 | 0.01669 | 3305.4 | 3706.0 | 6.615 |
| 24 | 950 | 0.01682 | 3316.1 | 3719.7 | 6.629 |
| 24 | 955 | 0.01694 | 3326.8 | 3733.4 | 6.643 |
| 24 | 960 | 0.01707 | 3337.4 | 3747.1 | 6.658 |
| 24 | 965 | 0.01719 | 3348.1 | 3760.7 | 6.672 |
| 24 | 970 | 0.01731 | 3358.7 | 3774.2 | 6.686 |
| 24 | 975 | 0.01744 | 3369.3 | 3787.8 | 6.700 |
| 24 | 980 | 0.01756 | 3379.9 | 3801.3 | 6.714 |
| 24 | 985 | 0.01768 | 3390.4 | 3814.7 | 6.727 |
| 24 | 990 | 0.01780 | 3401.0 | 3828.2 | 6.741 |
| 24 | 995 | 0.01792 | 3411.5 | 3841.6 | 6.754 |

| p | t | v | u | h | s |
|---|---|---|---|---|---|
| 24 | 1000 | 0.01804 | 3422.0 | 3855.0 | 6.768 |
| 24 | 1005 | 0.01816 | 3432.5 | 3868.4 | 6.781 |
| 24 | 1010 | 0.01828 | 3443.0 | 3881.7 | 6.794 |
| 24 | 1015 | 0.01840 | 3453.5 | 3895.0 | 6.808 |
| 24 | 1020 | 0.01851 | 3464.0 | 3908.3 | 6.821 |
| 24 | 1025 | 0.01863 | 3474.4 | 3921.6 | 6.834 |
| 24 | 1030 | 0.01875 | 3484.9 | 3934.9 | 6.847 |
| 24 | 1035 | 0.01887 | 3495.3 | 3948.1 | 6.859 |
| 24 | 1040 | 0.01898 | 3505.8 | 3961.4 | 6.872 |
| 24 | 1045 | 0.01910 | 3516.2 | 3974.6 | 6.885 |
| 24 | 1050 | 0.01921 | 3526.6 | 3987.8 | 6.897 |
| 24 | 1055 | 0.01933 | 3537.1 | 4001.0 | 6.910 |
| 24 | 1060 | 0.01945 | 3547.5 | 4014.2 | 6.922 |
| 24 | 1065 | 0.01956 | 3557.9 | 4027.3 | 6.935 |
| 24 | 1070 | 0.01967 | 3568.3 | 4040.5 | 6.947 |

Pressure, p = 25 Mpa

| p | t | v | u | h | s |
|---|---|---|---|---|---|
| 25 | 275 | 0.0009881 | 7.864 | 32.57 | 0.02815 |
| 25 | 280 | 0.0009884 | 28.40 | 53.11 | 0.1022 |

| p | t | v | u | h | s |
|---|---|---|---|---|---|
| 25 | 285 | 0.0009891 | 48.94 | 73.66 | 0.1749 |
| 25 | 290 | 0.0009901 | 69.47 | 94.22 | 0.2465 |
| 25 | 295 | 0.0009912 | 90.01 | 114.8 | 0.3168 |
| 25 | 300 | 0.0009926 | 110.6 | 135.4 | 0.3860 |
| 25 | 305 | 0.0009942 | 131.1 | 156.0 | 0.4540 |
| 25 | 310 | 0.0009960 | 151.7 | 176.6 | 0.5210 |
| 25 | 315 | 0.0009979 | 172.2 | 197.2 | 0.5870 |
| 25 | 320 | 0.001000 | 192.8 | 217.8 | 0.6519 |
| 25 | 325 | 0.001002 | 213.4 | 238.4 | 0.7159 |
| 25 | 330 | 0.001005 | 233.9 | 259.1 | 0.7789 |
| 25 | 335 | 0.001007 | 254.5 | 279.7 | 0.8410 |
| 25 | 340 | 0.001010 | 275.1 | 300.4 | 0.9023 |
| 25 | 345 | 0.001013 | 295.7 | 321.1 | 0.9626 |
| 25 | 350 | 0.001016 | 316.4 | 341.8 | 1.022 |
| 25 | 355 | 0.001019 | 337.0 | 362.5 | 1.081 |
| 25 | 360 | 0.001022 | 357.7 | 383.2 | 1.139 |
| 25 | 365 | 0.001026 | 378.4 | 404.0 | 1.196 |
| 25 | 370 | 0.001029 | 399.0 | 424.8 | 1.253 |
| 25 | 375 | 0.001033 | 419.8 | 445.6 | 1.309 |
| 25 | 380 | 0.001036 | 440.5 | 466.4 | 1.364 |

| p | t | v | u | h | s |
|---|---|---|---|---|---|
| 25 | 385 | 0.001040 | 461.3 | 487.3 | 1.419 |
| 25 | 390 | 0.001044 | 482.1 | 508.2 | 1.472 |
| 25 | 395 | 0.001049 | 502.9 | 529.1 | 1.526 |
| 25 | 400 | 0.001053 | 523.8 | 550.1 | 1.579 |
| 25 | 405 | 0.001057 | 544.7 | 571.1 | 1.631 |
| 25 | 410 | 0.001062 | 565.6 | 592.2 | 1.682 |
| 25 | 415 | 0.001067 | 586.6 | 613.3 | 1.734 |
| 25 | 420 | 0.001072 | 607.6 | 634.4 | 1.784 |
| 25 | 425 | 0.001077 | 628.7 | 655.6 | 1.834 |
| 25 | 430 | 0.001082 | 649.8 | 676.8 | 1.884 |
| 25 | 435 | 0.001087 | 670.9 | 698.1 | 1.933 |
| 25 | 440 | 0.001093 | 692.1 | 719.5 | 1.982 |
| 25 | 445 | 0.001099 | 713.4 | 740.9 | 2.030 |
| 25 | 450 | 0.001105 | 734.7 | 762.3 | 2.078 |
| 25 | 455 | 0.001111 | 756.1 | 783.9 | 2.126 |
| 25 | 460 | 0.001117 | 777.6 | 805.5 | 2.173 |
| 25 | 465 | 0.001124 | 799.1 | 827.2 | 2.220 |
| 25 | 470 | 0.001130 | 820.7 | 849.0 | 2.267 |
| 25 | 475 | 0.001137 | 842.4 | 870.8 | 2.313 |
| 25 | 480 | 0.001144 | 864.2 | 892.8 | 2.359 |

| p | t | v | u | h | s |
|---|---|---|---|---|---|
| 25 | 485 | 0.001152 | 886.1 | 914.9 | 2.405 |
| 25 | 490 | 0.001160 | 908.0 | 937.0 | 2.450 |
| 25 | 495 | 0.001167 | 930.1 | 959.3 | 2.495 |
| 25 | 500 | 0.001176 | 952.3 | 981.7 | 2.540 |
| 25 | 505 | 0.001184 | 974.7 | 1004.3 | 2.585 |
| 25 | 510 | 0.001193 | 997.1 | 1026.9 | 2.630 |
| 25 | 515 | 0.001202 | 1019.7 | 1049.8 | 2.675 |
| 25 | 520 | 0.001212 | 1042.5 | 1072.8 | 2.719 |
| 25 | 525 | 0.001222 | 1065.4 | 1095.9 | 2.763 |
| 25 | 530 | 0.001232 | 1088.5 | 1119.3 | 2.808 |
| 25 | 535 | 0.001243 | 1111.7 | 1142.8 | 2.852 |
| 25 | 540 | 0.001254 | 1135.2 | 1166.6 | 2.896 |
| 25 | 545 | 0.001266 | 1158.9 | 1190.6 | 2.940 |
| 25 | 550 | 0.001279 | 1182.9 | 1214.8 | 2.985 |
| 25 | 555 | 0.001292 | 1207.0 | 1239.3 | 3.029 |
| 25 | 560 | 0.001306 | 1231.5 | 1264.2 | 3.073 |
| 25 | 565 | 0.001320 | 1256.3 | 1289.3 | 3.118 |
| 25 | 570 | 0.001336 | 1281.4 | 1314.8 | 3.163 |
| 25 | 575 | 0.001352 | 1306.9 | 1340.7 | 3.208 |
| 25 | 580 | 0.001369 | 1332.8 | 1367.0 | 3.254 |

| p | t | v | u | h | s |
|---|---|---|---|---|---|
| 25 | 585 | 0.001388 | 1359.1 | 1393.8 | 3.300 |
| 25 | 590 | 0.001408 | 1386.0 | 1421.2 | 3.346 |
| 25 | 595 | 0.001430 | 1413.4 | 1449.1 | 3.394 |
| 25 | 600 | 0.001453 | 1441.5 | 1477.8 | 3.442 |
| 25 | 605 | 0.001479 | 1470.3 | 1507.3 | 3.491 |
| 25 | 610 | 0.001507 | 1500.1 | 1537.7 | 3.541 |
| 25 | 615 | 0.001538 | 1530.8 | 1569.3 | 3.592 |
| 25 | 620 | 0.001574 | 1562.9 | 1602.3 | 3.646 |
| 25 | 625 | 0.001615 | 1596.5 | 1636.9 | 3.701 |
| 25 | 630 | 0.001662 | 1632.3 | 1673.8 | 3.760 |
| 25 | 635 | 0.001720 | 1670.9 | 1713.9 | 3.823 |
| 25 | 640 | 0.001792 | 1713.6 | 1758.4 | 3.893 |
| 25 | 645 | 0.001891 | 1763.0 | 1810.2 | 3.974 |
| 25 | 650 | 0.002046 | 1825.2 | 1876.4 | 4.076 |
| 25 | 655 | 0.002400 | 1927.5 | 1987.5 | 4.246 |
| 25 | 660 | 0.003867 | 2179.5 | 2276.2 | 4.685 |
| 25 | 665 | 0.004977 | 2317.9 | 2442.4 | 4.936 |
| 25 | 670 | 0.005660 | 2392.8 | 2534.3 | 5.074 |
| 25 | 675 | 0.006188 | 2447.0 | 2601.7 | 5.174 |
| 25 | 680 | 0.006633 | 2490.9 | 2656.7 | 5.255 |

Thermodynamic Properties of Supercritical Steam

| p | t | v | u | h | s |
|---|---|---|---|---|---|
| 25 | 685 | 0.007023 | 2528.2 | 2703.8 | 5.324 |
| 25 | 690 | 0.007374 | 2561.1 | 2745.4 | 5.385 |
| 25 | 695 | 0.007695 | 2590.5 | 2782.9 | 5.439 |
| 25 | 700 | 0.007994 | 2617.5 | 2817.3 | 5.488 |
| 25 | 705 | 0.008273 | 2642.4 | 2849.2 | 5.534 |
| 25 | 710 | 0.008538 | 2665.7 | 2879.1 | 5.576 |
| 25 | 715 | 0.008789 | 2687.6 | 2907.3 | 5.616 |
| 25 | 720 | 0.009029 | 2708.4 | 2934.1 | 5.653 |
| 25 | 725 | 0.009259 | 2728.2 | 2959.7 | 5.688 |
| 25 | 730 | 0.009481 | 2747.2 | 2984.2 | 5.722 |
| 25 | 735 | 0.009695 | 2765.4 | 3007.8 | 5.754 |
| 25 | 740 | 0.009902 | 2783.0 | 3030.6 | 5.785 |
| 25 | 745 | 0.01010 | 2800.0 | 3052.6 | 5.815 |
| 25 | 750 | 0.01030 | 2816.5 | 3074.0 | 5.843 |
| 25 | 755 | 0.01049 | 2832.5 | 3094.7 | 5.871 |
| 25 | 760 | 0.01067 | 2848.1 | 3114.9 | 5.898 |
| 25 | 765 | 0.01085 | 2863.3 | 3134.7 | 5.924 |
| 25 | 770 | 0.01103 | 2878.2 | 3154.0 | 5.949 |
| 25 | 775 | 0.01121 | 2892.7 | 3172.9 | 5.973 |
| 25 | 780 | 0.01138 | 2907.0 | 3191.4 | 5.997 |

| p | t | v | u | h | s |
|---|---|---|---|---|---|
| 25 | 785 | 0.01154 | 2920.9 | 3209.5 | 6.020 |
| 25 | 790 | 0.01171 | 2934.7 | 3227.4 | 6.043 |
| 25 | 795 | 0.01187 | 2948.2 | 3244.9 | 6.065 |
| 25 | 800 | 0.01203 | 2961.5 | 3262.2 | 6.087 |
| 25 | 805 | 0.01218 | 2974.7 | 3279.3 | 6.108 |
| 25 | 810 | 0.01234 | 2987.6 | 3296.1 | 6.129 |
| 25 | 815 | 0.01249 | 3000.4 | 3312.7 | 6.149 |
| 25 | 820 | 0.01264 | 3013.0 | 3329.1 | 6.169 |
| 25 | 825 | 0.01279 | 3025.5 | 3345.3 | 6.189 |
| 25 | 830 | 0.01294 | 3037.9 | 3361.3 | 6.208 |
| 25 | 835 | 0.01308 | 3050.1 | 3377.1 | 6.227 |
| 25 | 840 | 0.01322 | 3062.2 | 3392.8 | 6.246 |
| 25 | 845 | 0.01337 | 3074.2 | 3408.4 | 6.265 |
| 25 | 850 | 0.01351 | 3086.2 | 3423.8 | 6.283 |
| 25 | 855 | 0.01365 | 3098.0 | 3439.1 | 6.301 |
| 25 | 860 | 0.01378 | 3109.7 | 3454.3 | 6.318 |
| 25 | 865 | 0.01392 | 3121.4 | 3469.4 | 6.336 |
| 25 | 870 | 0.01406 | 3132.9 | 3484.3 | 6.353 |
| 25 | 875 | 0.01419 | 3144.4 | 3499.2 | 6.370 |
| 25 | 880 | 0.01432 | 3155.9 | 3513.9 | 6.387 |

Thermodynamic Properties of Supercritical Steam

| p | t | v | u | h | s |
|---|---|---|---|---|---|
| 25 | 885 | 0.01445 | 3167.2 | 3528.6 | 6.404 |
| 25 | 890 | 0.01458 | 3178.6 | 3543.2 | 6.420 |
| 25 | 895 | 0.01471 | 3189.8 | 3557.7 | 6.436 |
| 25 | 900 | 0.01484 | 3201.0 | 3572.1 | 6.452 |
| 25 | 905 | 0.01497 | 3212.2 | 3586.4 | 6.468 |
| 25 | 910 | 0.01510 | 3223.2 | 3600.7 | 6.484 |
| 25 | 915 | 0.01523 | 3234.3 | 3614.9 | 6.499 |
| 25 | 920 | 0.01535 | 3245.3 | 3629.1 | 6.515 |
| 25 | 925 | 0.01548 | 3256.3 | 3643.2 | 6.530 |
| 25 | 930 | 0.01560 | 3267.2 | 3657.2 | 6.545 |
| 25 | 935 | 0.01572 | 3278.1 | 3671.2 | 6.560 |
| 25 | 940 | 0.01585 | 3289.0 | 3685.1 | 6.575 |
| 25 | 945 | 0.01597 | 3299.8 | 3699.0 | 6.590 |
| 25 | 950 | 0.01609 | 3310.6 | 3712.8 | 6.605 |
| 25 | 955 | 0.01621 | 3321.4 | 3726.6 | 6.619 |
| 25 | 960 | 0.01633 | 3332.1 | 3740.4 | 6.633 |
| 25 | 965 | 0.01645 | 3342.9 | 3754.1 | 6.648 |
| 25 | 970 | 0.01657 | 3353.6 | 3767.8 | 6.662 |
| 25 | 975 | 0.01669 | 3364.2 | 3781.4 | 6.676 |
| 25 | 980 | 0.01681 | 3374.9 | 3795.0 | 6.690 |

| p | t | v | u | h | s |
|---|---|---|---|---|---|
| 25 | 985 | 0.01692 | 3385.5 | 3808.6 | 6.703 |
| 25 | 990 | 0.01704 | 3396.1 | 3822.1 | 6.717 |
| 25 | 995 | 0.01716 | 3406.7 | 3835.6 | 6.731 |
| 25 | 1000 | 0.01727 | 3417.3 | 3849.1 | 6.744 |
| 25 | 1005 | 0.01739 | 3427.9 | 3862.6 | 6.758 |
| 25 | 1010 | 0.01750 | 3438.5 | 3876.0 | 6.771 |
| 25 | 1015 | 0.01762 | 3449.0 | 3889.4 | 6.784 |
| 25 | 1020 | 0.01773 | 3459.5 | 3902.8 | 6.797 |
| 25 | 1025 | 0.01784 | 3470.1 | 3916.2 | 6.811 |
| 25 | 1030 | 0.01796 | 3480.6 | 3929.5 | 6.824 |
| 25 | 1035 | 0.01807 | 3491.1 | 3942.8 | 6.836 |
| 25 | 1040 | 0.01818 | 3501.6 | 3956.2 | 6.849 |
| 25 | 1045 | 0.01830 | 3512.0 | 3969.4 | 6.862 |
| 25 | 1050 | 0.01841 | 3522.5 | 3982.7 | 6.875 |
| 25 | 1055 | 0.01852 | 3533.0 | 3996.0 | 6.887 |
| 25 | 1060 | 0.01863 | 3543.4 | 4009.2 | 6.900 |
| 25 | 1065 | 0.01874 | 3553.9 | 4022.5 | 6.912 |
| 25 | 1070 | 0.01885 | 3564.4 | 4035.7 | 6.925 |

| p | t | v | u | h | s |
|---|---|---|---|---|---|

Pressure, p= 27 Mpa

| p | t | v | u | h | s |
|---|---|---|---|---|---|
| 27 | 275 | 0.0009871 | 7.861 | 34.51 | 0.02805 |
| 27 | 280 | 0.0009875 | 28.36 | 55.03 | 0.1020 |
| 27 | 285 | 0.0009882 | 48.86 | 75.55 | 0.1746 |
| 27 | 290 | 0.0009892 | 69.37 | 96.08 | 0.2460 |
| 27 | 295 | 0.0009904 | 89.88 | 116.6 | 0.3163 |
| 27 | 300 | 0.0009918 | 110.4 | 137.2 | 0.3854 |
| 27 | 305 | 0.0009934 | 130.9 | 157.7 | 0.4534 |
| 27 | 310 | 0.0009951 | 151.5 | 178.3 | 0.5203 |
| 27 | 315 | 0.0009971 | 172.0 | 198.9 | 0.5862 |
| 27 | 320 | 0.0009992 | 192.5 | 219.5 | 0.6510 |
| 27 | 325 | 0.001001 | 213.1 | 240.1 | 0.7149 |
| 27 | 330 | 0.001004 | 233.6 | 260.7 | 0.7779 |
| 27 | 335 | 0.001006 | 254.2 | 281.4 | 0.8400 |
| 27 | 340 | 0.001009 | 274.8 | 302.0 | 0.9011 |
| 27 | 345 | 0.001012 | 295.4 | 322.7 | 0.9615 |
| 27 | 350 | 0.001015 | 316.0 | 343.4 | 1.021 |
| 27 | 355 | 0.001018 | 336.6 | 364.1 | 1.080 |
| 27 | 360 | 0.001021 | 357.2 | 384.8 | 1.138 |
| 27 | 365 | 0.001025 | 377.9 | 405.5 | 1.195 |

| p | t | v | u | h | s |
|---|---|---|---|---|---|
| 27 | 370 | 0.001028 | 398.5 | 426.3 | 1.251 |
| 27 | 375 | 0.001032 | 419.2 | 447.1 | 1.307 |
| 27 | 380 | 0.001036 | 440.0 | 467.9 | 1.362 |
| 27 | 385 | 0.001039 | 460.7 | 488.8 | 1.417 |
| 27 | 390 | 0.001043 | 481.5 | 509.6 | 1.471 |
| 27 | 395 | 0.001048 | 502.3 | 530.6 | 1.524 |
| 27 | 400 | 0.001052 | 523.1 | 551.5 | 1.577 |
| 27 | 405 | 0.001056 | 544.0 | 572.5 | 1.629 |
| 27 | 410 | 0.001061 | 564.9 | 593.5 | 1.681 |
| 27 | 415 | 0.001066 | 585.8 | 614.6 | 1.732 |
| 27 | 420 | 0.001071 | 606.8 | 635.7 | 1.782 |
| 27 | 425 | 0.001076 | 627.8 | 656.9 | 1.832 |
| 27 | 430 | 0.001081 | 648.9 | 678.1 | 1.882 |
| 27 | 435 | 0.001086 | 670.0 | 699.3 | 1.931 |
| 27 | 440 | 0.001092 | 691.2 | 720.6 | 1.980 |
| 27 | 445 | 0.001097 | 712.4 | 742.0 | 2.028 |
| 27 | 450 | 0.001103 | 733.7 | 763.5 | 2.076 |
| 27 | 455 | 0.001109 | 755.0 | 785.0 | 2.124 |
| 27 | 460 | 0.001116 | 776.4 | 806.6 | 2.171 |
| 27 | 465 | 0.001122 | 797.9 | 828.2 | 2.217 |

| p | t | v | u | h | s |
|---|---|---|---|---|---|
| 27 | 470 | 0.001129 | 819.5 | 849.9 | 2.264 |
| 27 | 475 | 0.001136 | 841.1 | 871.8 | 2.310 |
| 27 | 480 | 0.001143 | 862.8 | 893.7 | 2.356 |
| 27 | 485 | 0.001150 | 884.7 | 915.7 | 2.402 |
| 27 | 490 | 0.001158 | 906.6 | 937.8 | 2.447 |
| 27 | 495 | 0.001165 | 928.6 | 960.1 | 2.492 |
| 27 | 500 | 0.001174 | 950.7 | 982.4 | 2.537 |
| 27 | 505 | 0.001182 | 973.0 | 1004.9 | 2.582 |
| 27 | 510 | 0.001191 | 995.3 | 1027.5 | 2.626 |
| 27 | 515 | 0.001200 | 1017.9 | 1050.3 | 2.671 |
| 27 | 520 | 0.001209 | 1040.5 | 1073.2 | 2.715 |
| 27 | 525 | 0.001219 | 1063.3 | 1096.2 | 2.759 |
| 27 | 530 | 0.001229 | 1086.3 | 1119.5 | 2.803 |
| 27 | 535 | 0.001240 | 1109.5 | 1143.0 | 2.847 |
| 27 | 540 | 0.001251 | 1132.8 | 1166.6 | 2.891 |
| 27 | 545 | 0.001263 | 1156.4 | 1190.5 | 2.935 |
| 27 | 550 | 0.001275 | 1180.2 | 1214.6 | 2.979 |
| 27 | 555 | 0.001288 | 1204.2 | 1239.0 | 3.024 |
| 27 | 560 | 0.001301 | 1228.5 | 1263.6 | 3.068 |
| 27 | 565 | 0.001315 | 1253.1 | 1288.6 | 3.112 |

| p | t | v | u | h | s |
|---|---|---|---|---|---|
| 27 | 570 | 0.001330 | 1278.0 | 1313.9 | 3.157 |
| 27 | 575 | 0.001346 | 1303.2 | 1339.6 | 3.202 |
| 27 | 580 | 0.001363 | 1328.8 | 1365.6 | 3.247 |
| 27 | 585 | 0.001381 | 1354.9 | 1392.2 | 3.292 |
| 27 | 590 | 0.001400 | 1381.4 | 1419.2 | 3.338 |
| 27 | 595 | 0.001421 | 1408.4 | 1446.8 | 3.385 |
| 27 | 600 | 0.001443 | 1436.0 | 1475.0 | 3.432 |
| 27 | 605 | 0.001468 | 1464.3 | 1503.9 | 3.480 |
| 27 | 610 | 0.001494 | 1493.4 | 1533.7 | 3.529 |
| 27 | 615 | 0.001523 | 1523.3 | 1564.5 | 3.579 |
| 27 | 620 | 0.001556 | 1554.4 | 1596.4 | 3.631 |
| 27 | 625 | 0.001593 | 1586.7 | 1629.8 | 3.685 |
| 27 | 630 | 0.001635 | 1620.7 | 1664.9 | 3.741 |
| 27 | 635 | 0.001685 | 1656.9 | 1702.4 | 3.800 |
| 27 | 640 | 0.001745 | 1695.9 | 1743.0 | 3.864 |
| 27 | 645 | 0.001820 | 1739.0 | 1788.1 | 3.934 |
| 27 | 650 | 0.001921 | 1788.3 | 1840.2 | 4.014 |
| 27 | 655 | 0.002073 | 1849.0 | 1905.0 | 4.114 |
| 27 | 660 | 0.002358 | 1935.6 | 1999.3 | 4.257 |
| 27 | 665 | 0.003080 | 2087.8 | 2170.9 | 4.516 |

Thermodynamic Properties of Supercritical Steam

| p | t | v | u | h | s |
|---|---|---|---|---|---|
| 27 | 670 | 0.004082 | 2242.7 | 2353.0 | 4.789 |
| 27 | 675 | 0.004808 | 2336.5 | 2466.3 | 4.957 |
| 27 | 680 | 0.005358 | 2401.3 | 2545.9 | 5.075 |
| 27 | 685 | 0.005813 | 2451.9 | 2608.9 | 5.167 |
| 27 | 690 | 0.006208 | 2494.2 | 2661.8 | 5.244 |
| 27 | 695 | 0.006559 | 2530.8 | 2707.9 | 5.311 |
| 27 | 700 | 0.006879 | 2563.3 | 2749.0 | 5.370 |
| 27 | 705 | 0.007174 | 2592.7 | 2786.4 | 5.423 |
| 27 | 710 | 0.007449 | 2619.6 | 2820.8 | 5.471 |
| 27 | 715 | 0.007708 | 2644.7 | 2852.8 | 5.516 |
| 27 | 720 | 0.007953 | 2668.2 | 2882.9 | 5.558 |
| 27 | 725 | 0.008186 | 2690.3 | 2911.3 | 5.598 |
| 27 | 730 | 0.008410 | 2711.3 | 2938.4 | 5.635 |
| 27 | 735 | 0.008625 | 2731.4 | 2964.3 | 5.670 |
| 27 | 740 | 0.008831 | 2750.6 | 2989.1 | 5.704 |
| 27 | 745 | 0.009031 | 2769.1 | 3013.0 | 5.736 |
| 27 | 750 | 0.009225 | 2786.9 | 3036.0 | 5.767 |
| 27 | 755 | 0.009413 | 2804.2 | 3058.3 | 5.796 |
| 27 | 760 | 0.009595 | 2820.9 | 3080.0 | 5.825 |
| 27 | 765 | 0.009773 | 2837.1 | 3101.0 | 5.853 |

| p | t | v | u | h | s |
|---|---|---|---|---|---|
| 27 | 770 | 0.009947 | 2852.9 | 3121.5 | 5.879 |
| 27 | 775 | 0.01012 | 2868.4 | 3141.5 | 5.905 |
| 27 | 780 | 0.01028 | 2883.4 | 3161.1 | 5.930 |
| 27 | 785 | 0.01045 | 2898.2 | 3180.2 | 5.955 |
| 27 | 790 | 0.01060 | 2912.7 | 3199.0 | 5.979 |
| 27 | 795 | 0.01076 | 2926.9 | 3217.4 | 6.002 |
| 27 | 800 | 0.01092 | 2940.8 | 3235.5 | 6.025 |
| 27 | 805 | 0.01107 | 2954.5 | 3253.3 | 6.047 |
| 27 | 810 | 0.01122 | 2968.0 | 3270.8 | 6.069 |
| 27 | 815 | 0.01136 | 2981.3 | 3288.1 | 6.090 |
| 27 | 820 | 0.01151 | 2994.5 | 3305.2 | 6.111 |
| 27 | 825 | 0.01165 | 3007.4 | 3322.0 | 6.131 |
| 27 | 830 | 0.01179 | 3020.2 | 3338.6 | 6.151 |
| 27 | 835 | 0.01193 | 3032.9 | 3355.0 | 6.171 |
| 27 | 840 | 0.01207 | 3045.4 | 3371.2 | 6.190 |
| 27 | 845 | 0.01220 | 3057.8 | 3387.3 | 6.209 |
| 27 | 850 | 0.01234 | 3070.1 | 3403.2 | 6.228 |
| 27 | 855 | 0.01247 | 3082.3 | 3419.0 | 6.247 |
| 27 | 860 | 0.01260 | 3094.4 | 3434.6 | 6.265 |
| 27 | 865 | 0.01273 | 3106.4 | 3450.1 | 6.283 |

Thermodynamic Properties of Supercritical Steam

| p | t | v | u | h | s |
|---|---|---|---|---|---|
| 27 | 870 | 0.01286 | 3118.2 | 3465.5 | 6.301 |
| 27 | 875 | 0.01299 | 3130.0 | 3480.7 | 6.318 |
| 27 | 880 | 0.01312 | 3141.8 | 3495.9 | 6.335 |
| 27 | 885 | 0.01324 | 3153.4 | 3510.9 | 6.352 |
| 27 | 890 | 0.01337 | 3165.0 | 3525.8 | 6.369 |
| 27 | 895 | 0.01349 | 3176.5 | 3540.7 | 6.386 |
| 27 | 900 | 0.01361 | 3187.9 | 3555.4 | 6.402 |
| 27 | 905 | 0.01373 | 3199.3 | 3570.1 | 6.418 |
| 27 | 910 | 0.01385 | 3210.7 | 3584.7 | 6.435 |
| 27 | 915 | 0.01397 | 3221.9 | 3599.2 | 6.450 |
| 27 | 920 | 0.01409 | 3233.2 | 3613.7 | 6.466 |
| 27 | 925 | 0.01421 | 3244.4 | 3628.0 | 6.482 |
| 27 | 930 | 0.01433 | 3255.5 | 3642.3 | 6.497 |
| 27 | 935 | 0.01444 | 3266.6 | 3656.6 | 6.512 |
| 27 | 940 | 0.01456 | 3277.6 | 3670.8 | 6.528 |
| 27 | 945 | 0.01468 | 3288.7 | 3684.9 | 6.543 |
| 27 | 950 | 0.01479 | 3299.6 | 3699.0 | 6.557 |
| 27 | 955 | 0.01491 | 3310.6 | 3713.0 | 6.572 |
| 27 | 960 | 0.01502 | 3321.5 | 3727.0 | 6.587 |
| 27 | 965 | 0.01513 | 3332.4 | 3740.9 | 6.601 |

| p | t | v | u | h | s |
|---|---|---|---|---|---|
| 27 | 970 | 0.01524 | 3343.2 | 3754.8 | 6.616 |
| 27 | 975 | 0.01536 | 3354.1 | 3768.7 | 6.630 |
| 27 | 980 | 0.01547 | 3364.9 | 3782.5 | 6.644 |
| 27 | 985 | 0.01558 | 3375.7 | 3796.3 | 6.658 |
| 27 | 990 | 0.01569 | 3386.4 | 3810.0 | 6.672 |
| 27 | 995 | 0.01580 | 3397.2 | 3823.7 | 6.686 |
| 27 | 1000 | 0.01591 | 3407.9 | 3837.4 | 6.699 |
| 27 | 1005 | 0.01602 | 3418.6 | 3851.0 | 6.713 |
| 27 | 1010 | 0.01612 | 3429.3 | 3864.6 | 6.727 |
| 27 | 1015 | 0.01623 | 3439.9 | 3878.2 | 6.740 |
| 27 | 1020 | 0.01634 | 3450.6 | 3891.7 | 6.753 |
| 27 | 1025 | 0.01645 | 3461.2 | 3905.3 | 6.766 |
| 27 | 1030 | 0.01655 | 3471.9 | 3918.8 | 6.780 |
| 27 | 1035 | 0.01666 | 3482.5 | 3932.3 | 6.793 |
| 27 | 1040 | 0.01676 | 3493.1 | 3945.7 | 6.806 |
| 27 | 1045 | 0.01687 | 3503.7 | 3959.2 | 6.819 |
| 27 | 1050 | 0.01698 | 3514.2 | 3972.6 | 6.831 |
| 27 | 1055 | 0.01708 | 3524.8 | 3986.0 | 6.844 |
| 27 | 1060 | 0.01718 | 3535.4 | 3999.4 | 6.857 |
| 27 | 1065 | 0.01729 | 3545.9 | 4012.7 | 6.869 |

| p | t | v | u | h | s |
|---|---|---|---|---|---|
| 27 | 1070 | 0.01739 | 3556.5 | 4026.1 | 6.882 |

Pressure, p = 30 Mpa

| p | t | v | u | h | s |
|---|---|---|---|---|---|
| 30 | 275 | 0.0009858 | 7.854 | 37.43 | 0.02789 |
| 30 | 280 | 0.0009862 | 28.30 | 57.89 | 0.1016 |
| 30 | 285 | 0.0009869 | 48.76 | 78.36 | 0.1741 |
| 30 | 290 | 0.0009879 | 69.22 | 98.86 | 0.2454 |
| 30 | 295 | 0.0009891 | 89.69 | 119.4 | 0.3155 |
| 30 | 300 | 0.0009905 | 110.2 | 139.9 | 0.3845 |
| 30 | 305 | 0.0009921 | 130.7 | 160.4 | 0.4524 |
| 30 | 310 | 0.0009939 | 151.1 | 181.0 | 0.5192 |
| 30 | 315 | 0.0009958 | 171.6 | 201.5 | 0.5849 |
| 30 | 320 | 0.0009979 | 192.1 | 222.1 | 0.6497 |
| 30 | 325 | 0.001000 | 212.7 | 242.7 | 0.7135 |
| 30 | 330 | 0.001003 | 233.2 | 263.3 | 0.7764 |
| 30 | 335 | 0.001005 | 253.7 | 283.9 | 0.8384 |
| 30 | 340 | 0.001008 | 274.2 | 304.5 | 0.8995 |
| 30 | 345 | 0.001011 | 294.8 | 325.1 | 0.9597 |
| 30 | 350 | 0.001014 | 315.4 | 345.8 | 1.019 |
| 30 | 355 | 0.001017 | 335.9 | 366.4 | 1.078 |

| p | t | v | u | h | s |
|---|---|---|---|---|---|
| 30 | 360 | 0.001020 | 356.5 | 387.1 | 1.136 |
| 30 | 365 | 0.001023 | 377.2 | 407.9 | 1.193 |
| 30 | 370 | 0.001027 | 397.8 | 428.6 | 1.249 |
| 30 | 375 | 0.001030 | 418.4 | 449.4 | 1.305 |
| 30 | 380 | 0.001034 | 439.1 | 470.1 | 1.360 |
| 30 | 385 | 0.001038 | 459.8 | 491.0 | 1.415 |
| 30 | 390 | 0.001042 | 480.6 | 511.8 | 1.468 |
| 30 | 395 | 0.001046 | 501.3 | 532.7 | 1.522 |
| 30 | 400 | 0.001050 | 522.1 | 553.6 | 1.574 |
| 30 | 405 | 0.001055 | 542.9 | 574.6 | 1.626 |
| 30 | 410 | 0.001059 | 563.8 | 595.6 | 1.678 |
| 30 | 415 | 0.001064 | 584.7 | 616.6 | 1.729 |
| 30 | 420 | 0.001069 | 605.6 | 637.7 | 1.779 |
| 30 | 425 | 0.001074 | 626.6 | 658.8 | 1.829 |
| 30 | 430 | 0.001079 | 647.6 | 680.0 | 1.879 |
| 30 | 435 | 0.001084 | 668.6 | 701.2 | 1.928 |
| 30 | 440 | 0.001090 | 689.8 | 722.4 | 1.976 |
| 30 | 445 | 0.001095 | 710.9 | 743.8 | 2.025 |
| 30 | 450 | 0.001101 | 732.1 | 765.2 | 2.072 |
| 30 | 455 | 0.001107 | 753.4 | 786.6 | 2.120 |

Thermodynamic Properties of Supercritical Steam

| p | t | v | u | h | s |
|---|---|---|---|---|---|
| 30 | 460 | 0.001113 | 774.8 | 808.2 | 2.167 |
| 30 | 465 | 0.001120 | 796.2 | 829.8 | 2.214 |
| 30 | 470 | 0.001126 | 817.7 | 851.4 | 2.260 |
| 30 | 475 | 0.001133 | 839.2 | 873.2 | 2.306 |
| 30 | 480 | 0.001140 | 860.9 | 895.1 | 2.352 |
| 30 | 485 | 0.001147 | 882.6 | 917.0 | 2.397 |
| 30 | 490 | 0.001155 | 904.4 | 939.0 | 2.442 |
| 30 | 495 | 0.001162 | 926.3 | 961.2 | 2.487 |
| 30 | 500 | 0.001170 | 948.4 | 983.5 | 2.532 |
| 30 | 505 | 0.001179 | 970.5 | 1005.8 | 2.577 |
| 30 | 510 | 0.001187 | 992.7 | 1028.4 | 2.621 |
| 30 | 515 | 0.001196 | 1015.1 | 1051.0 | 2.665 |
| 30 | 520 | 0.001205 | 1037.7 | 1073.8 | 2.709 |
| 30 | 525 | 0.001215 | 1060.3 | 1096.8 | 2.753 |
| 30 | 530 | 0.001225 | 1083.2 | 1119.9 | 2.797 |
| 30 | 535 | 0.001235 | 1106.2 | 1143.2 | 2.841 |
| 30 | 540 | 0.001246 | 1129.3 | 1166.7 | 2.885 |
| 30 | 545 | 0.001257 | 1152.7 | 1190.4 | 2.928 |
| 30 | 550 | 0.001269 | 1176.3 | 1214.4 | 2.972 |
| 30 | 555 | 0.001282 | 1200.1 | 1238.5 | 3.016 |

| p | t | v | u | h | s |
|---|---|---|---|---|---|
| 30 | 560 | 0.001295 | 1224.1 | 1263.0 | 3.060 |
| 30 | 565 | 0.001308 | 1248.4 | 1287.7 | 3.104 |
| 30 | 570 | 0.001323 | 1273.0 | 1312.7 | 3.148 |
| 30 | 575 | 0.001338 | 1297.9 | 1338.1 | 3.192 |
| 30 | 580 | 0.001354 | 1323.2 | 1363.8 | 3.237 |
| 30 | 585 | 0.001371 | 1348.8 | 1389.9 | 3.281 |
| 30 | 590 | 0.001389 | 1374.8 | 1416.5 | 3.327 |
| 30 | 595 | 0.001409 | 1401.3 | 1443.6 | 3.372 |
| 30 | 600 | 0.001430 | 1428.3 | 1471.2 | 3.419 |
| 30 | 605 | 0.001452 | 1455.9 | 1499.5 | 3.466 |
| 30 | 610 | 0.001477 | 1484.1 | 1528.4 | 3.513 |
| 30 | 615 | 0.001503 | 1513.1 | 1558.2 | 3.562 |
| 30 | 620 | 0.001533 | 1542.9 | 1588.9 | 3.611 |
| 30 | 625 | 0.001565 | 1573.7 | 1620.7 | 3.663 |
| 30 | 630 | 0.001602 | 1605.8 | 1653.8 | 3.715 |
| 30 | 635 | 0.001644 | 1639.4 | 1688.7 | 3.770 |
| 30 | 640 | 0.001692 | 1674.8 | 1725.6 | 3.828 |
| 30 | 645 | 0.001750 | 1712.7 | 1765.2 | 3.890 |
| 30 | 650 | 0.001820 | 1753.9 | 1808.5 | 3.957 |
| 30 | 655 | 0.001909 | 1799.7 | 1856.9 | 4.031 |

Thermodynamic Properties of Supercritical Steam

| p | t | v | u | h | s |
|---|---|---|---|---|---|
| 30 | 660 | 0.002030 | 1852.4 | 1913.3 | 4.117 |
| 30 | 665 | 0.002209 | 1916.9 | 1983.2 | 4.222 |
| 30 | 670 | 0.002507 | 2001.5 | 2076.7 | 4.362 |
| 30 | 675 | 0.003005 | 2111.0 | 2201.1 | 4.547 |
| 30 | 680 | 0.003620 | 2219.8 | 2328.4 | 4.735 |
| 30 | 685 | 0.004181 | 2305.1 | 2430.6 | 4.885 |
| 30 | 690 | 0.004656 | 2370.7 | 2510.4 | 5.001 |
| 30 | 695 | 0.005066 | 2423.7 | 2575.7 | 5.095 |
| 30 | 700 | 0.005429 | 2468.6 | 2631.5 | 5.175 |
| 30 | 705 | 0.005755 | 2507.5 | 2680.1 | 5.245 |
| 30 | 710 | 0.006053 | 2542.0 | 2723.6 | 5.306 |
| 30 | 715 | 0.006328 | 2573.2 | 2763.1 | 5.361 |
| 30 | 720 | 0.006584 | 2601.8 | 2799.4 | 5.412 |
| 30 | 725 | 0.006826 | 2628.4 | 2833.1 | 5.459 |
| 30 | 730 | 0.007055 | 2653.2 | 2864.8 | 5.502 |
| 30 | 735 | 0.007273 | 2676.5 | 2894.7 | 5.543 |
| 30 | 740 | 0.007481 | 2698.7 | 2923.1 | 5.582 |
| 30 | 745 | 0.007682 | 2719.8 | 2950.2 | 5.618 |
| 30 | 750 | 0.007874 | 2739.9 | 2976.2 | 5.653 |
| 30 | 755 | 0.008061 | 2759.3 | 3001.1 | 5.686 |

| p | t | v | u | h | s |
|---|---|---|---|---|---|
| 30 | 760 | 0.008241 | 2777.9 | 3025.2 | 5.718 |
| 30 | 765 | 0.008416 | 2795.9 | 3048.4 | 5.748 |
| 30 | 770 | 0.008586 | 2813.4 | 3070.9 | 5.778 |
| 30 | 775 | 0.00875 | 2830.3 | 3092.8 | 5.806 |
| 30 | 780 | 0.00891 | 2846.7 | 3114.1 | 5.833 |
| 30 | 785 | 0.00907 | 2862.8 | 3134.9 | 5.860 |
| 30 | 790 | 0.00922 | 2878.4 | 3155.1 | 5.886 |
| 30 | 795 | 0.00938 | 2893.7 | 3175.0 | 5.911 |
| 30 | 800 | 0.00952 | 2908.7 | 3194.4 | 5.935 |
| 30 | 805 | 0.00967 | 2923.4 | 3213.4 | 5.959 |
| 30 | 810 | 0.00981 | 2937.8 | 3232.1 | 5.982 |
| 30 | 815 | 0.00995 | 2951.9 | 3250.5 | 6.004 |
| 30 | 820 | 0.01009 | 2965.9 | 3268.5 | 6.027 |
| 30 | 825 | 0.01022 | 2979.6 | 3286.3 | 6.048 |
| 30 | 830 | 0.01036 | 2993.1 | 3303.9 | 6.069 |
| 30 | 835 | 0.01049 | 3006.5 | 3321.2 | 6.090 |
| 30 | 840 | 0.01062 | 3019.7 | 3338.3 | 6.111 |
| 30 | 845 | 0.01075 | 3032.7 | 3355.2 | 6.131 |
| 30 | 850 | 0.01088 | 3045.6 | 3371.8 | 6.150 |
| 30 | 855 | 0.01100 | 3058.3 | 3388.3 | 6.170 |

Thermodynamic Properties of Supercritical Steam

| p | t | v | u | h | s |
|---|---|---|---|---|---|
| 30 | 860 | 0.01112 | 3070.9 | 3404.7 | 6.189 |
| 30 | 865 | 0.01125 | 3083.4 | 3420.8 | 6.207 |
| 30 | 870 | 0.01137 | 3095.8 | 3436.9 | 6.226 |
| 30 | 875 | 0.01149 | 3108.1 | 3452.7 | 6.244 |
| 30 | 880 | 0.01161 | 3120.3 | 3468.5 | 6.262 |
| 30 | 885 | 0.01172 | 3132.3 | 3484.1 | 6.280 |
| 30 | 890 | 0.01184 | 3144.3 | 3499.6 | 6.297 |
| 30 | 895 | 0.01196 | 3156.2 | 3515.0 | 6.314 |
| 30 | 900 | 0.01207 | 3168.1 | 3530.2 | 6.331 |
| 30 | 905 | 0.01218 | 3179.8 | 3545.4 | 6.348 |
| 30 | 910 | 0.01230 | 3191.5 | 3560.5 | 6.365 |
| 30 | 915 | 0.01241 | 3203.2 | 3575.4 | 6.381 |
| 30 | 920 | 0.01252 | 3214.7 | 3590.3 | 6.398 |
| 30 | 925 | 0.01263 | 3226.2 | 3605.1 | 6.414 |
| 30 | 930 | 0.01274 | 3237.7 | 3619.9 | 6.429 |
| 30 | 935 | 0.01285 | 3249.1 | 3634.5 | 6.445 |
| 30 | 940 | 0.01296 | 3260.4 | 3649.1 | 6.461 |
| 30 | 945 | 0.01306 | 3271.7 | 3663.6 | 6.476 |
| 30 | 950 | 0.01317 | 3283.0 | 3678.1 | 6.491 |
| 30 | 955 | 0.01328 | 3294.2 | 3692.5 | 6.506 |

| p | t | v | u | h | s |
|---|---|---|---|---|---|
| 30 | 960 | 0.01338 | 3305.4 | 3706.8 | 6.521 |
| 30 | 965 | 0.01349 | 3316.5 | 3721.1 | 6.536 |
| 30 | 970 | 0.01359 | 3327.6 | 3735.3 | 6.551 |
| 30 | 975 | 0.01369 | 3338.7 | 3749.5 | 6.566 |
| 30 | 980 | 0.01380 | 3349.7 | 3763.6 | 6.580 |
| 30 | 985 | 0.01390 | 3360.8 | 3777.7 | 6.594 |
| 30 | 990 | 0.01400 | 3371.7 | 3791.7 | 6.609 |
| 30 | 995 | 0.01410 | 3382.7 | 3805.7 | 6.623 |
| 30 | 1000 | 0.01420 | 3393.6 | 3819.7 | 6.637 |
| 30 | 1005 | 0.01430 | 3404.5 | 3833.6 | 6.651 |
| 30 | 1010 | 0.01440 | 3415.4 | 3847.4 | 6.664 |
| 30 | 1015 | 0.01450 | 3426.3 | 3861.3 | 6.678 |
| 30 | 1020 | 0.01460 | 3437.1 | 3875.1 | 6.692 |
| 30 | 1025 | 0.01470 | 3447.9 | 3888.9 | 6.705 |
| 30 | 1030 | 0.01480 | 3458.7 | 3902.6 | 6.718 |
| 30 | 1035 | 0.01489 | 3469.5 | 3916.3 | 6.732 |
| 30 | 1040 | 0.01499 | 3480.3 | 3930.0 | 6.745 |
| 30 | 1045 | 0.01509 | 3491.0 | 3943.7 | 6.758 |
| 30 | 1050 | 0.01519 | 3501.8 | 3957.4 | 6.771 |
| 30 | 1055 | 0.01528 | 3512.5 | 3971.0 | 6.784 |

| p | t | v | u | h | s |
|---|---|---|---|---|---|
| 30 | 1060 | 0.01538 | 3523.2 | 3984.6 | 6.797 |
| 30 | 1065 | 0.01547 | 3533.9 | 3998.1 | 6.810 |
| 30 | 1070 | 0.01557 | 3544.6 | 4011.7 | 6.822 |

Pressure, p = 35 Mpa

| | | | | | |
|---|---|---|---|---|---|
| 35 | 275 | 0.0009835 | 7.836 | 42.26 | 0.02755 |
| 35 | 280 | 0.0009840 | 28.20 | 62.64 | 0.1010 |
| 35 | 285 | 0.0009848 | 48.57 | 83.04 | 0.1732 |
| 35 | 290 | 0.0009858 | 68.97 | 103.4 | 0.2443 |
| 35 | 295 | 0.0009870 | 89.37 | 123.9 | 0.3142 |
| 35 | 300 | 0.0009885 | 109.8 | 144.4 | 0.3830 |
| 35 | 305 | 0.0009901 | 130.2 | 164.9 | 0.4507 |
| 35 | 310 | 0.0009919 | 150.6 | 185.3 | 0.5173 |
| 35 | 315 | 0.0009938 | 171.1 | 205.9 | 0.5829 |
| 35 | 320 | 0.0009959 | 191.5 | 226.4 | 0.6476 |
| 35 | 325 | 0.0009982 | 212.0 | 246.9 | 0.7112 |
| 35 | 330 | 0.001001 | 232.4 | 267.5 | 0.7740 |
| 35 | 335 | 0.001003 | 252.9 | 288.0 | 0.8358 |
| 35 | 340 | 0.001006 | 273.4 | 308.6 | 0.8967 |

| p | t | v | u | h | s |
|---|---|---|---|---|---|
| 35 | 345 | 0.001009 | 293.9 | 329.2 | 0.9569 |
| 35 | 350 | 0.001011 | 314.4 | 349.8 | 1.016 |
| 35 | 355 | 0.001015 | 334.9 | 370.4 | 1.075 |
| 35 | 360 | 0.001018 | 355.4 | 391.1 | 1.132 |
| 35 | 365 | 0.001021 | 376.0 | 411.7 | 1.189 |
| 35 | 370 | 0.001024 | 396.6 | 432.4 | 1.246 |
| 35 | 375 | 0.001028 | 417.2 | 453.1 | 1.301 |
| 35 | 380 | 0.001032 | 437.8 | 473.9 | 1.356 |
| 35 | 385 | 0.001036 | 458.4 | 494.6 | 1.411 |
| 35 | 390 | 0.001040 | 479.1 | 515.4 | 1.464 |
| 35 | 395 | 0.001044 | 499.7 | 536.3 | 1.517 |
| 35 | 400 | 0.001048 | 520.5 | 557.1 | 1.570 |
| 35 | 405 | 0.001052 | 541.2 | 578.0 | 1.622 |
| 35 | 410 | 0.001057 | 562.0 | 599.0 | 1.673 |
| 35 | 415 | 0.001061 | 582.8 | 619.9 | 1.724 |
| 35 | 420 | 0.001066 | 603.6 | 641.0 | 1.774 |
| 35 | 425 | 0.001071 | 624.5 | 662.0 | 1.824 |
| 35 | 430 | 0.001076 | 645.5 | 683.1 | 1.874 |
| 35 | 435 | 0.001081 | 666.4 | 704.3 | 1.922 |
| 35 | 440 | 0.001087 | 687.4 | 725.5 | 1.971 |

Thermodynamic Properties of Supercritical Steam

| p | t | v | u | h | s |
|---|---|---|---|---|---|
| 35 | 445 | 0.001092 | 708.5 | 746.7 | 2.019 |
| 35 | 450 | 0.001098 | 729.6 | 768.0 | 2.067 |
| 35 | 455 | 0.001104 | 750.8 | 789.4 | 2.114 |
| 35 | 460 | 0.001110 | 772.0 | 810.9 | 2.161 |
| 35 | 465 | 0.001116 | 793.3 | 832.4 | 2.207 |
| 35 | 470 | 0.001122 | 814.7 | 854.0 | 2.253 |
| 35 | 475 | 0.001129 | 836.1 | 875.6 | 2.299 |
| 35 | 480 | 0.001136 | 857.6 | 897.4 | 2.345 |
| 35 | 485 | 0.001143 | 879.2 | 919.2 | 2.390 |
| 35 | 490 | 0.001150 | 900.9 | 941.1 | 2.435 |
| 35 | 495 | 0.001157 | 922.7 | 963.2 | 2.480 |
| 35 | 500 | 0.001165 | 944.5 | 985.3 | 2.524 |
| 35 | 505 | 0.001173 | 966.5 | 1007.5 | 2.568 |
| 35 | 510 | 0.001182 | 988.6 | 1029.9 | 2.613 |
| 35 | 515 | 0.001190 | 1010.7 | 1052.4 | 2.656 |
| 35 | 520 | 0.001199 | 1033.1 | 1075.0 | 2.700 |
| 35 | 525 | 0.001208 | 1055.5 | 1097.8 | 2.744 |
| 35 | 530 | 0.001218 | 1078.1 | 1120.7 | 2.787 |
| 35 | 535 | 0.001228 | 1100.8 | 1143.8 | 2.831 |
| 35 | 540 | 0.001238 | 1123.7 | 1167.1 | 2.874 |

| p | t | v | u | h | s |
|---|---|---|---|---|---|
| 35 | 545 | 0.001249 | 1146.8 | 1190.5 | 2.917 |
| 35 | 550 | 0.001260 | 1170.1 | 1214.2 | 2.960 |
| 35 | 555 | 0.001272 | 1193.5 | 1238.1 | 3.004 |
| 35 | 560 | 0.001284 | 1217.2 | 1262.2 | 3.047 |
| 35 | 565 | 0.001297 | 1241.1 | 1286.5 | 3.090 |
| 35 | 570 | 0.001311 | 1265.3 | 1311.1 | 3.133 |
| 35 | 575 | 0.001325 | 1289.7 | 1336.1 | 3.177 |
| 35 | 580 | 0.001340 | 1314.4 | 1361.3 | 3.221 |
| 35 | 585 | 0.001356 | 1339.4 | 1386.9 | 3.265 |
| 35 | 590 | 0.001373 | 1364.8 | 1412.8 | 3.309 |
| 35 | 595 | 0.001390 | 1390.5 | 1439.2 | 3.353 |
| 35 | 600 | 0.001409 | 1416.6 | 1466.0 | 3.398 |
| 35 | 605 | 0.001430 | 1443.2 | 1493.3 | 3.443 |
| 35 | 610 | 0.001451 | 1470.3 | 1521.1 | 3.489 |
| 35 | 615 | 0.001475 | 1498.0 | 1549.6 | 3.536 |
| 35 | 620 | 0.001500 | 1526.2 | 1578.8 | 3.583 |
| 35 | 625 | 0.001528 | 1555.2 | 1608.7 | 3.631 |
| 35 | 630 | 0.001558 | 1585.1 | 1639.6 | 3.680 |
| 35 | 635 | 0.001592 | 1615.9 | 1671.6 | 3.731 |
| 35 | 640 | 0.001629 | 1647.8 | 1704.8 | 3.783 |

| p | t | v | u | h | s |
|---|---|---|---|---|---|
| 35 | 645 | 0.001672 | 1681.1 | 1739.6 | 3.837 |
| 35 | 650 | 0.001720 | 1716.1 | 1776.3 | 3.894 |
| 35 | 655 | 0.001777 | 1753.0 | 1815.2 | 3.953 |
| 35 | 660 | 0.001844 | 1792.5 | 1857.0 | 4.017 |
| 35 | 665 | 0.001925 | 1835.3 | 1902.6 | 4.086 |
| 35 | 670 | 0.002027 | 1882.3 | 1953.2 | 4.162 |
| 35 | 675 | 0.002159 | 1935.0 | 2010.6 | 4.247 |
| 35 | 680 | 0.002334 | 1995.0 | 2076.7 | 4.344 |
| 35 | 685 | 0.002569 | 2063.1 | 2153.0 | 4.456 |
| 35 | 690 | 0.002868 | 2136.8 | 2237.2 | 4.579 |
| 35 | 695 | 0.003212 | 2210.1 | 2322.5 | 4.702 |
| 35 | 700 | 0.003565 | 2277.2 | 2402.0 | 4.816 |
| 35 | 705 | 0.003906 | 2336.3 | 2473.0 | 4.917 |
| 35 | 710 | 0.004224 | 2387.9 | 2535.7 | 5.006 |
| 35 | 715 | 0.004520 | 2433.4 | 2591.6 | 5.084 |
| 35 | 720 | 0.004796 | 2474.0 | 2641.8 | 5.154 |
| 35 | 725 | 0.005052 | 2510.5 | 2687.3 | 5.217 |
| 35 | 730 | 0.005292 | 2543.7 | 2729.0 | 5.274 |
| 35 | 735 | 0.005518 | 2574.4 | 2767.5 | 5.327 |
| 35 | 740 | 0.005731 | 2602.8 | 2803.4 | 5.376 |

| p | t | v | u | h | s |
|---|---|---|---|---|---|
| 35 | 745 | 0.005935 | 2629.3 | 2837.1 | 5.421 |
| 35 | 750 | 0.006129 | 2654.4 | 2868.9 | 5.463 |
| 35 | 755 | 0.006314 | 2678.1 | 2899.1 | 5.504 |
| 35 | 760 | 0.006493 | 2700.6 | 2927.8 | 5.542 |
| 35 | 765 | 0.006665 | 2722.1 | 2955.4 | 5.578 |
| 35 | 770 | 0.006831 | 2742.8 | 2981.9 | 5.612 |
| 35 | 775 | 0.006992 | 2762.6 | 3007.3 | 5.645 |
| 35 | 780 | 0.007149 | 2781.7 | 3031.9 | 5.677 |
| 35 | 785 | 0.007301 | 2800.2 | 3055.8 | 5.707 |
| 35 | 790 | 0.007448 | 2818.2 | 3078.9 | 5.737 |
| 35 | 795 | 0.007593 | 2835.6 | 3101.3 | 5.765 |
| 35 | 800 | 0.007733 | 2852.5 | 3123.2 | 5.792 |
| 35 | 805 | 0.007871 | 2869.0 | 3144.5 | 5.819 |
| 35 | 810 | 0.008006 | 2885.1 | 3165.3 | 5.845 |
| 35 | 815 | 0.008138 | 2900.9 | 3185.7 | 5.870 |
| 35 | 820 | 0.008267 | 2916.3 | 3205.7 | 5.894 |
| 35 | 825 | 0.008394 | 2931.5 | 3225.3 | 5.918 |
| 35 | 830 | 0.008519 | 2946.3 | 3244.5 | 5.941 |
| 35 | 835 | 0.008642 | 2960.9 | 3263.4 | 5.964 |
| 35 | 840 | 0.008763 | 2975.3 | 3282.0 | 5.986 |

Thermodynamic Properties of Supercritical Steam

| p | t | v | u | h | s |
|---|---|---|---|---|---|
| 35 | 845 | 0.008882 | 2989.5 | 3300.3 | 6.008 |
| 35 | 850 | 0.009000 | 3003.4 | 3318.4 | 6.029 |
| 35 | 855 | 0.009115 | 3017.2 | 3336.2 | 6.050 |
| 35 | 860 | 0.009230 | 3030.7 | 3353.8 | 6.071 |
| 35 | 865 | 0.009342 | 3044.2 | 3371.1 | 6.091 |
| 35 | 870 | 0.009454 | 3057.4 | 3388.3 | 6.111 |
| 35 | 875 | 0.009564 | 3070.5 | 3405.3 | 6.130 |
| 35 | 880 | 0.009672 | 3083.5 | 3422.1 | 6.149 |
| 35 | 885 | 0.009780 | 3096.4 | 3438.7 | 6.168 |
| 35 | 890 | 0.009886 | 3109.1 | 3455.2 | 6.187 |
| 35 | 895 | 0.00999 | 3121.8 | 3471.5 | 6.205 |
| 35 | 900 | 0.01010 | 3134.3 | 3487.6 | 6.223 |
| 35 | 905 | 0.01020 | 3146.7 | 3503.7 | 6.241 |
| 35 | 910 | 0.01030 | 3159.0 | 3519.6 | 6.258 |
| 35 | 915 | 0.01040 | 3171.3 | 3535.4 | 6.275 |
| 35 | 920 | 0.01050 | 3183.5 | 3551.1 | 6.293 |
| 35 | 925 | 0.01060 | 3195.5 | 3566.6 | 6.309 |
| 35 | 930 | 0.01070 | 3207.5 | 3582.1 | 6.326 |
| 35 | 935 | 0.01080 | 3219.5 | 3597.4 | 6.343 |
| 35 | 940 | 0.01090 | 3231.3 | 3612.7 | 6.359 |

| p | t | v | u | h | s |
|---|---|---|---|---|---|
| 35 | 945 | 0.01099 | 3243.1 | 3627.9 | 6.375 |
| 35 | 950 | 0.01109 | 3254.9 | 3643.0 | 6.391 |
| 35 | 955 | 0.01118 | 3266.6 | 3658.0 | 6.407 |
| 35 | 960 | 0.01128 | 3278.2 | 3672.9 | 6.422 |
| 35 | 965 | 0.01137 | 3289.8 | 3687.8 | 6.438 |
| 35 | 970 | 0.01147 | 3301.3 | 3702.6 | 6.453 |
| 35 | 975 | 0.01156 | 3312.8 | 3717.3 | 6.468 |
| 35 | 980 | 0.01165 | 3324.2 | 3732.0 | 6.483 |
| 35 | 985 | 0.01174 | 3335.6 | 3746.6 | 6.498 |
| 35 | 990 | 0.01183 | 3347.0 | 3761.1 | 6.513 |
| 35 | 995 | 0.01192 | 3358.3 | 3775.6 | 6.527 |
| 35 | 1000 | 0.01201 | 3369.6 | 3790.1 | 6.542 |
| 35 | 1005 | 0.01210 | 3380.9 | 3804.4 | 6.556 |
| 35 | 1010 | 0.01219 | 3392.1 | 3818.8 | 6.570 |
| 35 | 1015 | 0.01228 | 3403.3 | 3833.1 | 6.584 |
| 35 | 1020 | 0.01237 | 3414.4 | 3847.3 | 6.598 |
| 35 | 1025 | 0.01246 | 3425.6 | 3861.5 | 6.612 |
| 35 | 1030 | 0.01254 | 3436.7 | 3875.7 | 6.626 |
| 35 | 1035 | 0.01263 | 3447.8 | 3889.8 | 6.640 |
| 35 | 1040 | 0.01272 | 3458.8 | 3903.9 | 6.653 |

Thermodynamic Properties of Supercritical Steam

| p | t | v | u | h | s |
|---|---|---|---|---|---|
| 35 | 1045 | 0.01280 | 3469.9 | 3917.9 | 6.667 |
| 35 | 1050 | 0.01289 | 3480.9 | 3932.0 | 6.680 |
| 35 | 1055 | 0.01297 | 3491.9 | 3945.9 | 6.694 |
| 35 | 1060 | 0.01306 | 3502.9 | 3959.9 | 6.707 |
| 35 | 1065 | 0.01314 | 3513.8 | 3973.8 | 6.720 |
| 35 | 1070 | 0.01323 | 3524.8 | 3987.7 | 6.733 |

Pressure, p = 40 Mpa

| p | t | v | u | h | s |
|---|---|---|---|---|---|
| 40 | 275 | 0.0009813 | 7.808 | 47.06 | 0.02715 |
| 40 | 280 | 0.0009818 | 28.09 | 67.36 | 0.1003 |
| 40 | 285 | 0.0009827 | 48.39 | 87.70 | 0.1723 |
| 40 | 290 | 0.0009837 | 68.71 | 108.1 | 0.2431 |
| 40 | 295 | 0.0009850 | 89.05 | 128.4 | 0.3128 |
| 40 | 300 | 0.0009864 | 109.4 | 148.9 | 0.3814 |
| 40 | 305 | 0.0009881 | 129.8 | 169.3 | 0.4490 |
| 40 | 310 | 0.0009899 | 150.1 | 189.7 | 0.5154 |
| 40 | 315 | 0.0009918 | 170.5 | 210.2 | 0.5809 |
| 40 | 320 | 0.0009939 | 190.9 | 230.7 | 0.6454 |
| 40 | 325 | 0.0009962 | 211.3 | 251.1 | 0.7089 |
| 40 | 330 | 0.0009985 | 231.7 | 271.6 | 0.7715 |

| p | t | v | u | h | s |
|---|---|---|---|---|---|
| 40 | 335 | 0.001001 | 252.1 | 292.2 | 0.8332 |
| 40 | 340 | 0.001004 | 272.5 | 312.7 | 0.8940 |
| 40 | 345 | 0.001006 | 293.0 | 333.2 | 0.9540 |
| 40 | 350 | 0.001009 | 313.4 | 353.8 | 1.013 |
| 40 | 355 | 0.001012 | 333.9 | 374.4 | 1.072 |
| 40 | 360 | 0.001016 | 354.4 | 395.0 | 1.129 |
| 40 | 365 | 0.001019 | 374.8 | 415.6 | 1.186 |
| 40 | 370 | 0.001022 | 395.4 | 436.2 | 1.242 |
| 40 | 375 | 0.001026 | 415.9 | 456.9 | 1.298 |
| 40 | 380 | 0.001029 | 436.4 | 477.6 | 1.353 |
| 40 | 385 | 0.001033 | 457.0 | 498.3 | 1.407 |
| 40 | 390 | 0.001037 | 477.6 | 519.1 | 1.460 |
| 40 | 395 | 0.001041 | 498.2 | 539.9 | 1.513 |
| 40 | 400 | 0.001045 | 518.9 | 560.7 | 1.566 |
| 40 | 405 | 0.001050 | 539.5 | 581.5 | 1.617 |
| 40 | 410 | 0.001054 | 560.2 | 602.4 | 1.669 |
| 40 | 415 | 0.001059 | 581.0 | 623.3 | 1.719 |
| 40 | 420 | 0.001063 | 601.7 | 644.3 | 1.770 |
| 40 | 425 | 0.001068 | 622.5 | 665.3 | 1.819 |
| 40 | 430 | 0.001073 | 643.4 | 686.3 | 1.868 |

| p | t | v | u | h | s |
|---|---|---|---|---|---|
| 40 | 435 | 0.001078 | 664.3 | 707.4 | 1.917 |
| 40 | 440 | 0.001083 | 685.2 | 728.5 | 1.966 |
| 40 | 445 | 0.001089 | 706.2 | 749.7 | 2.013 |
| 40 | 450 | 0.001094 | 727.2 | 771.0 | 2.061 |
| 40 | 455 | 0.001100 | 748.3 | 792.3 | 2.108 |
| 40 | 460 | 0.001106 | 769.4 | 813.6 | 2.155 |
| 40 | 465 | 0.001112 | 790.6 | 835.0 | 2.201 |
| 40 | 470 | 0.001118 | 811.8 | 856.5 | 2.247 |
| 40 | 475 | 0.001125 | 833.1 | 878.1 | 2.293 |
| 40 | 480 | 0.001131 | 854.5 | 899.8 | 2.338 |
| 40 | 485 | 0.001138 | 876.0 | 921.5 | 2.383 |
| 40 | 490 | 0.001145 | 897.5 | 943.3 | 2.428 |
| 40 | 495 | 0.001153 | 919.1 | 965.2 | 2.472 |
| 40 | 500 | 0.001160 | 940.8 | 987.2 | 2.516 |
| 40 | 505 | 0.001168 | 962.6 | 1009.3 | 2.560 |
| 40 | 510 | 0.001176 | 984.5 | 1031.6 | 2.604 |
| 40 | 515 | 0.001184 | 1006.5 | 1053.9 | 2.648 |
| 40 | 520 | 0.001193 | 1028.6 | 1076.4 | 2.691 |
| 40 | 525 | 0.001202 | 1050.9 | 1099.0 | 2.734 |
| 40 | 530 | 0.001211 | 1073.3 | 1121.7 | 2.778 |

| p | t | v | u | h | s |
|---|---|---|---|---|---|
| 40 | 535 | 0.001221 | 1095.8 | 1144.6 | 2.821 |
| 40 | 540 | 0.001231 | 1118.4 | 1167.6 | 2.863 |
| 40 | 545 | 0.001241 | 1141.2 | 1190.9 | 2.906 |
| 40 | 550 | 0.001252 | 1164.2 | 1214.3 | 2.949 |
| 40 | 555 | 0.001263 | 1187.4 | 1237.9 | 2.992 |
| 40 | 560 | 0.001275 | 1210.7 | 1261.7 | 3.034 |
| 40 | 565 | 0.001287 | 1234.2 | 1285.7 | 3.077 |
| 40 | 570 | 0.001300 | 1258.0 | 1310.0 | 3.120 |
| 40 | 575 | 0.001313 | 1282.0 | 1334.5 | 3.163 |
| 40 | 580 | 0.001327 | 1306.2 | 1359.3 | 3.206 |
| 40 | 585 | 0.001342 | 1330.7 | 1384.4 | 3.249 |
| 40 | 590 | 0.001358 | 1355.5 | 1409.8 | 3.292 |
| 40 | 595 | 0.001374 | 1380.6 | 1435.6 | 3.336 |
| 40 | 600 | 0.001392 | 1406.1 | 1461.7 | 3.379 |
| 40 | 605 | 0.001410 | 1431.9 | 1488.3 | 3.423 |
| 40 | 610 | 0.001430 | 1458.1 | 1515.3 | 3.468 |
| 40 | 615 | 0.001451 | 1484.7 | 1542.8 | 3.513 |
| 40 | 620 | 0.001473 | 1511.8 | 1570.8 | 3.558 |
| 40 | 625 | 0.001498 | 1539.5 | 1599.4 | 3.604 |
| 40 | 630 | 0.001524 | 1567.8 | 1628.7 | 3.651 |

| p | t | v | u | h | s |
|---|---|---|---|---|---|
| 40 | 635 | 0.001552 | 1596.8 | 1658.9 | 3.698 |
| 40 | 640 | 0.001584 | 1626.5 | 1689.9 | 3.747 |
| 40 | 645 | 0.001618 | 1657.2 | 1721.9 | 3.797 |
| 40 | 650 | 0.001656 | 1688.9 | 1755.1 | 3.848 |
| 40 | 655 | 0.001699 | 1721.7 | 1789.7 | 3.901 |
| 40 | 660 | 0.001747 | 1756.0 | 1825.9 | 3.956 |
| 40 | 665 | 0.001802 | 1791.9 | 1863.9 | 4.014 |
| 40 | 670 | 0.001865 | 1829.7 | 1904.3 | 4.074 |
| 40 | 675 | 0.001940 | 1869.9 | 1947.4 | 4.138 |
| 40 | 680 | 0.002028 | 1912.8 | 1993.9 | 4.207 |
| 40 | 685 | 0.002135 | 1958.9 | 2044.3 | 4.281 |
| 40 | 690 | 0.002265 | 2008.7 | 2099.3 | 4.361 |
| 40 | 695 | 0.002423 | 2062.0 | 2158.9 | 4.447 |
| 40 | 700 | 0.002610 | 2118.1 | 2222.5 | 4.538 |
| 40 | 705 | 0.002824 | 2175.3 | 2288.3 | 4.632 |
| 40 | 710 | 0.003058 | 2231.6 | 2353.9 | 4.724 |
| 40 | 715 | 0.003300 | 2285.0 | 2417.0 | 4.813 |
| 40 | 720 | 0.003542 | 2334.8 | 2476.5 | 4.896 |
| 40 | 725 | 0.003780 | 2380.6 | 2531.7 | 4.972 |
| 40 | 730 | 0.004009 | 2422.6 | 2583.0 | 5.043 |

| p | t | v | u | h | s |
|---|---|---|---|---|---|
| 40 | 735 | 0.004229 | 2461.3 | 2630.4 | 5.108 |
| 40 | 740 | 0.004440 | 2496.9 | 2674.5 | 5.167 |
| 40 | 745 | 0.004641 | 2530.0 | 2715.6 | 5.223 |
| 40 | 750 | 0.004833 | 2560.8 | 2754.1 | 5.274 |
| 40 | 755 | 0.005017 | 2589.7 | 2790.3 | 5.322 |
| 40 | 760 | 0.005193 | 2616.8 | 2824.6 | 5.367 |
| 40 | 765 | 0.005362 | 2642.5 | 2857.0 | 5.410 |
| 40 | 770 | 0.005525 | 2666.9 | 2887.9 | 5.450 |
| 40 | 775 | 0.005682 | 2690.2 | 2917.5 | 5.489 |
| 40 | 780 | 0.005835 | 2712.4 | 2945.8 | 5.525 |
| 40 | 785 | 0.005982 | 2733.8 | 2973.0 | 5.560 |
| 40 | 790 | 0.006125 | 2754.3 | 2999.3 | 5.593 |
| 40 | 795 | 0.006264 | 2774.2 | 3024.7 | 5.625 |
| 40 | 800 | 0.006399 | 2793.3 | 3049.3 | 5.656 |
| 40 | 805 | 0.006531 | 2811.9 | 3073.2 | 5.686 |
| 40 | 810 | 0.006660 | 2830.0 | 3096.4 | 5.715 |
| 40 | 815 | 0.006786 | 2847.5 | 3119.0 | 5.742 |
| 40 | 820 | 0.006909 | 2864.7 | 3141.0 | 5.769 |
| 40 | 825 | 0.007030 | 2881.4 | 3162.6 | 5.795 |
| 40 | 830 | 0.007148 | 2897.7 | 3183.6 | 5.821 |

Thermodynamic Properties of Supercritical Steam

| p | t | v | u | h | s |
|---|---|---|---|---|---|
| 40 | 835 | 0.007264 | 2913.7 | 3204.3 | 5.846 |
| 40 | 840 | 0.007378 | 2929.4 | 3224.5 | 5.870 |
| 40 | 845 | 0.007490 | 2944.8 | 3244.4 | 5.893 |
| 40 | 850 | 0.007600 | 2959.9 | 3263.9 | 5.917 |
| 40 | 855 | 0.007708 | 2974.7 | 3283.1 | 5.939 |
| 40 | 860 | 0.007815 | 2989.4 | 3302.0 | 5.961 |
| 40 | 865 | 0.007921 | 3003.8 | 3320.6 | 5.983 |
| 40 | 870 | 0.008024 | 3018.0 | 3339.0 | 6.004 |
| 40 | 875 | 0.008127 | 3032.0 | 3357.1 | 6.025 |
| 40 | 880 | 0.008228 | 3045.9 | 3375.0 | 6.045 |
| 40 | 885 | 0.008328 | 3059.6 | 3392.7 | 6.065 |
| 40 | 890 | 0.008426 | 3073.1 | 3410.2 | 6.085 |
| 40 | 895 | 0.008524 | 3086.5 | 3427.5 | 6.104 |
| 40 | 900 | 0.008620 | 3099.8 | 3444.6 | 6.123 |
| 40 | 905 | 0.008716 | 3112.9 | 3461.5 | 6.142 |
| 40 | 910 | 0.008810 | 3125.9 | 3478.3 | 6.160 |
| 40 | 915 | 0.008903 | 3138.8 | 3495.0 | 6.179 |
| 40 | 920 | 0.008996 | 3151.6 | 3511.4 | 6.197 |
| 40 | 925 | 0.009088 | 3164.3 | 3527.8 | 6.214 |
| 40 | 930 | 0.009178 | 3176.9 | 3544.0 | 6.232 |

| p | t | v | u | h | s |
|---|---|---|---|---|---|
| 40 | 935 | 0.009268 | 3189.4 | 3560.1 | 6.249 |
| 40 | 940 | 0.009357 | 3201.8 | 3576.1 | 6.266 |
| 40 | 945 | 0.009446 | 3214.1 | 3592.0 | 6.283 |
| 40 | 950 | 0.009534 | 3226.4 | 3607.7 | 6.300 |
| 40 | 955 | 0.009621 | 3238.5 | 3623.4 | 6.316 |
| 40 | 960 | 0.009707 | 3250.7 | 3638.9 | 6.332 |
| 40 | 965 | 0.009793 | 3262.7 | 3654.4 | 6.348 |
| 40 | 970 | 0.009878 | 3274.7 | 3669.8 | 6.364 |
| 40 | 975 | 0.009962 | 3286.6 | 3685.1 | 6.380 |
| 40 | 980 | 0.01005 | 3298.4 | 3700.3 | 6.396 |
| 40 | 985 | 0.01013 | 3310.3 | 3715.4 | 6.411 |
| 40 | 990 | 0.01021 | 3322.0 | 3730.5 | 6.426 |
| 40 | 995 | 0.01029 | 3333.7 | 3745.5 | 6.441 |
| 40 | 1000 | 0.01038 | 3345.4 | 3760.4 | 6.456 |
| 40 | 1005 | 0.01046 | 3357.0 | 3775.3 | 6.471 |
| 40 | 1010 | 0.01054 | 3368.6 | 3790.1 | 6.486 |
| 40 | 1015 | 0.01062 | 3380.1 | 3804.8 | 6.500 |
| 40 | 1020 | 0.01070 | 3391.6 | 3819.5 | 6.515 |
| 40 | 1025 | 0.01078 | 3403.0 | 3834.2 | 6.529 |
| 40 | 1030 | 0.01086 | 3414.5 | 3848.7 | 6.543 |

Thermodynamic Properties of Supercritical Steam

| p | t | v | u | h | s |
|---|---|---|---|---|---|
| 40 | 1035 | 0.01094 | 3425.9 | 3863.3 | 6.557 |
| 40 | 1040 | 0.01101 | 3437.2 | 3877.8 | 6.571 |
| 40 | 1045 | 0.01109 | 3448.6 | 3892.2 | 6.585 |
| 40 | 1050 | 0.01117 | 3459.9 | 3906.6 | 6.599 |
| 40 | 1055 | 0.01125 | 3471.1 | 3921.0 | 6.613 |
| 40 | 1060 | 0.01132 | 3482.4 | 3935.3 | 6.626 |
| 40 | 1065 | 0.01140 | 3493.6 | 3949.6 | 6.640 |
| 40 | 1070 | 0.01148 | 3504.8 | 3963.8 | 6.653 |

Pressure, p = 45 Mpa

| p | t | v | u | h | s |
|---|---|---|---|---|---|
| 45 | 275 | 0.0009791 | 7.772 | 51.83 | 0.02668 |
| 45 | 280 | 0.0009797 | 27.97 | 72.06 | 0.09958 |
| 45 | 285 | 0.0009806 | 48.20 | 92.33 | 0.1713 |
| 45 | 290 | 0.0009817 | 68.46 | 112.6 | 0.2420 |
| 45 | 295 | 0.0009830 | 88.73 | 133.0 | 0.3115 |
| 45 | 300 | 0.0009844 | 109.0 | 153.3 | 0.3799 |
| 45 | 305 | 0.0009861 | 129.3 | 173.7 | 0.4472 |
| 45 | 310 | 0.0009879 | 149.6 | 194.1 | 0.5136 |
| 45 | 315 | 0.0009898 | 170.0 | 214.5 | 0.5789 |
| 45 | 320 | 0.0009919 | 190.3 | 234.9 | 0.6432 |

| p | t | v | u | h | s |
|---|---|---|---|---|---|
| 45 | 325 | 0.0009942 | 210.6 | 255.4 | 0.7066 |
| 45 | 330 | 0.0009966 | 231.0 | 275.8 | 0.7690 |
| 45 | 335 | 0.000999 | 251.3 | 296.3 | 0.8306 |
| 45 | 340 | 0.001002 | 271.7 | 316.8 | 0.8913 |
| 45 | 345 | 0.001004 | 292.1 | 337.3 | 0.9512 |
| 45 | 350 | 0.001007 | 312.5 | 357.8 | 1.010 |
| 45 | 355 | 0.001010 | 332.9 | 378.3 | 1.068 |
| 45 | 360 | 0.001013 | 353.3 | 398.9 | 1.126 |
| 45 | 365 | 0.001017 | 373.7 | 419.5 | 1.183 |
| 45 | 370 | 0.001020 | 394.2 | 440.1 | 1.239 |
| 45 | 375 | 0.001024 | 414.6 | 460.7 | 1.294 |
| 45 | 380 | 0.001027 | 435.1 | 481.3 | 1.349 |
| 45 | 385 | 0.001031 | 455.6 | 502.0 | 1.403 |
| 45 | 390 | 0.001035 | 476.2 | 522.7 | 1.456 |
| 45 | 395 | 0.001039 | 496.7 | 543.5 | 1.509 |
| 45 | 400 | 0.001043 | 517.3 | 564.2 | 1.561 |
| 45 | 405 | 0.001047 | 537.9 | 585.0 | 1.613 |
| 45 | 410 | 0.001051 | 558.5 | 605.8 | 1.664 |
| 45 | 415 | 0.001056 | 579.2 | 626.7 | 1.715 |
| 45 | 420 | 0.001061 | 599.9 | 647.6 | 1.765 |

Thermodynamic Properties of Supercritical Steam

| p | t | v | u | h | s |
|------|------|----------|--------|--------|-------|
| 45 | 425 | 0.001065 | 620.6 | 668.5 | 1.814 |
| 45 | 430 | 0.001070 | 641.4 | 689.5 | 1.863 |
| 45 | 435 | 0.001075 | 662.2 | 710.5 | 1.912 |
| 45 | 440 | 0.001080 | 683.0 | 731.6 | 1.960 |
| 45 | 445 | 0.001086 | 703.9 | 752.7 | 2.008 |
| 45 | 450 | 0.001091 | 724.8 | 773.9 | 2.055 |
| 45 | 455 | 0.001097 | 745.8 | 795.1 | 2.102 |
| 45 | 460 | 0.001103 | 766.8 | 816.4 | 2.149 |
| 45 | 465 | 0.001108 | 787.9 | 837.8 | 2.195 |
| 45 | 470 | 0.001115 | 809.0 | 859.2 | 2.241 |
| 45 | 475 | 0.001121 | 830.2 | 880.6 | 2.286 |
| 45 | 480 | 0.001127 | 851.5 | 902.2 | 2.331 |
| 45 | 485 | 0.001134 | 872.8 | 923.8 | 2.376 |
| 45 | 490 | 0.001141 | 894.2 | 945.5 | 2.421 |
| 45 | 495 | 0.001148 | 915.7 | 967.3 | 2.465 |
| 45 | 500 | 0.001156 | 937.2 | 989.2 | 2.509 |
| 45 | 505 | 0.001163 | 958.9 | 1011.2 | 2.553 |
| 45 | 510 | 0.001171 | 980.6 | 1033.3 | 2.596 |
| 45 | 515 | 0.001179 | 1002.5 | 1055.5 | 2.639 |
| 45 | 520 | 0.001187 | 1024.4 | 1077.8 | 2.683 |

| p | t | v | u | h | s |
|---|---|---|---|---|---|
| 45 | 525 | 0.001196 | 1046.4 | 1100.3 | 2.726 |
| 45 | 530 | 0.001205 | 1068.6 | 1122.8 | 2.768 |
| 45 | 535 | 0.001214 | 1090.9 | 1145.5 | 2.811 |
| 45 | 540 | 0.001224 | 1113.3 | 1168.4 | 2.854 |
| 45 | 545 | 0.001234 | 1135.9 | 1191.4 | 2.896 |
| 45 | 550 | 0.001244 | 1158.6 | 1214.6 | 2.938 |
| 45 | 555 | 0.001255 | 1181.5 | 1238.0 | 2.981 |
| 45 | 560 | 0.001266 | 1204.5 | 1261.5 | 3.023 |
| 45 | 565 | 0.001278 | 1227.7 | 1285.2 | 3.065 |
| 45 | 570 | 0.001290 | 1251.2 | 1309.2 | 3.107 |
| 45 | 575 | 0.001302 | 1274.8 | 1333.4 | 3.149 |
| 45 | 580 | 0.001316 | 1298.6 | 1357.8 | 3.192 |
| 45 | 585 | 0.001330 | 1322.7 | 1382.5 | 3.234 |
| 45 | 590 | 0.001344 | 1347.0 | 1407.5 | 3.277 |
| 45 | 595 | 0.001360 | 1371.5 | 1432.7 | 3.319 |
| 45 | 600 | 0.001376 | 1396.4 | 1458.3 | 3.362 |
| 45 | 605 | 0.001393 | 1421.6 | 1484.2 | 3.405 |
| 45 | 610 | 0.001411 | 1447.0 | 1510.5 | 3.448 |
| 45 | 615 | 0.001430 | 1472.9 | 1537.2 | 3.492 |
| 45 | 620 | 0.001450 | 1499.1 | 1564.4 | 3.536 |

Thermodynamic Properties of Supercritical Steam

| p | t | v | u | h | s |
|---|---|---|---|---|---|
| 45 | 625 | 0.001472 | 1525.8 | 1592.0 | 3.580 |
| 45 | 630 | 0.001495 | 1552.9 | 1620.2 | 3.625 |
| 45 | 635 | 0.001520 | 1580.5 | 1648.9 | 3.671 |
| 45 | 640 | 0.001547 | 1608.8 | 1678.4 | 3.717 |
| 45 | 645 | 0.001576 | 1637.6 | 1708.6 | 3.764 |
| 45 | 650 | 0.001608 | 1667.2 | 1739.6 | 3.812 |
| 45 | 655 | 0.001643 | 1697.6 | 1771.6 | 3.861 |
| 45 | 660 | 0.001681 | 1728.9 | 1804.6 | 3.911 |
| 45 | 665 | 0.001724 | 1761.2 | 1838.8 | 3.963 |
| 45 | 670 | 0.001771 | 1794.7 | 1874.4 | 4.016 |
| 45 | 675 | 0.001824 | 1829.4 | 1911.5 | 4.071 |
| 45 | 680 | 0.001884 | 1865.7 | 1950.5 | 4.129 |
| 45 | 685 | 0.001953 | 1903.6 | 1991.4 | 4.189 |
| 45 | 690 | 0.002032 | 1943.3 | 2034.7 | 4.252 |
| 45 | 695 | 0.002123 | 1985.0 | 2080.6 | 4.318 |
| 45 | 700 | 0.002228 | 2028.7 | 2129.0 | 4.387 |
| 45 | 705 | 0.002349 | 2074.3 | 2180.0 | 4.460 |
| 45 | 710 | 0.002486 | 2121.3 | 2233.2 | 4.535 |
| 45 | 715 | 0.002639 | 2169.1 | 2287.8 | 4.612 |
| 45 | 720 | 0.002805 | 2216.6 | 2342.9 | 4.688 |

| p | t | v | u | h | s |
|---|---|---|---|---|---|
| 45 | 725 | 0.002981 | 2263.2 | 2397.3 | 4.764 |
| 45 | 730 | 0.003162 | 2307.9 | 2450.3 | 4.837 |
| 45 | 735 | 0.003346 | 2350.6 | 2501.1 | 4.906 |
| 45 | 740 | 0.003528 | 2390.9 | 2549.7 | 4.972 |
| 45 | 745 | 0.003708 | 2428.8 | 2595.7 | 5.034 |
| 45 | 750 | 0.003884 | 2464.4 | 2639.2 | 5.092 |
| 45 | 755 | 0.004055 | 2497.9 | 2680.3 | 5.147 |
| 45 | 760 | 0.004221 | 2529.5 | 2719.5 | 5.198 |
| 45 | 765 | 0.004383 | 2559.4 | 2756.6 | 5.247 |
| 45 | 770 | 0.004539 | 2587.7 | 2792.0 | 5.293 |
| 45 | 775 | 0.004690 | 2614.6 | 2825.6 | 5.337 |
| 45 | 780 | 0.004836 | 2640.1 | 2857.7 | 5.378 |
| 45 | 785 | 0.004977 | 2664.5 | 2888.5 | 5.417 |
| 45 | 790 | 0.005115 | 2687.9 | 2918.0 | 5.455 |
| 45 | 795 | 0.005248 | 2710.3 | 2946.5 | 5.491 |
| 45 | 800 | 0.005378 | 2731.9 | 2973.9 | 5.525 |
| 45 | 805 | 0.005505 | 2752.7 | 3000.4 | 5.558 |
| 45 | 810 | 0.005628 | 2772.9 | 3026.1 | 5.590 |
| 45 | 815 | 0.005748 | 2792.4 | 3051.0 | 5.621 |
| 45 | 820 | 0.005866 | 2811.3 | 3075.3 | 5.650 |

| p | t | v | u | h | s |
|----|-----|----------|--------|--------|-------|
| 45 | 825 | 0.005981 | 2829.7 | 3098.8 | 5.679 |
| 45 | 830 | 0.006093 | 2847.6 | 3121.8 | 5.707 |
| 45 | 835 | 0.006203 | 2865.1 | 3144.3 | 5.734 |
| 45 | 840 | 0.006312 | 2882.2 | 3166.2 | 5.760 |
| 45 | 845 | 0.006418 | 2898.9 | 3187.7 | 5.785 |
| 45 | 850 | 0.006522 | 2915.3 | 3208.8 | 5.810 |
| 45 | 855 | 0.006624 | 2931.3 | 3229.4 | 5.835 |
| 45 | 860 | 0.006725 | 2947.1 | 3249.7 | 5.858 |
| 45 | 865 | 0.006824 | 2962.6 | 3269.7 | 5.881 |
| 45 | 870 | 0.006922 | 2977.8 | 3289.3 | 5.904 |
| 45 | 875 | 0.007018 | 2992.8 | 3308.6 | 5.926 |
| 45 | 880 | 0.007113 | 3007.6 | 3327.7 | 5.948 |
| 45 | 885 | 0.007207 | 3022.1 | 3346.4 | 5.969 |
| 45 | 890 | 0.007299 | 3036.5 | 3365.0 | 5.990 |
| 45 | 895 | 0.007390 | 3050.7 | 3383.3 | 6.010 |
| 45 | 900 | 0.007481 | 3064.7 | 3401.4 | 6.031 |
| 45 | 905 | 0.007570 | 3078.6 | 3419.2 | 6.050 |
| 45 | 910 | 0.007658 | 3092.3 | 3436.9 | 6.070 |
| 45 | 915 | 0.007745 | 3105.9 | 3454.4 | 6.089 |
| 45 | 920 | 0.007831 | 3119.3 | 3471.7 | 6.108 |

| p | t | v | u | h | s |
|---|---|---|---|---|---|
| 45 | 925 | 0.007916 | 3132.6 | 3488.9 | 6.127 |
| 45 | 930 | 0.008001 | 3145.8 | 3505.9 | 6.145 |
| 45 | 935 | 0.008084 | 3158.9 | 3522.7 | 6.163 |
| 45 | 940 | 0.008167 | 3171.9 | 3539.4 | 6.181 |
| 45 | 945 | 0.008249 | 3184.8 | 3556.0 | 6.198 |
| 45 | 950 | 0.008330 | 3197.6 | 3572.4 | 6.216 |
| 45 | 955 | 0.008411 | 3210.2 | 3588.7 | 6.233 |
| 45 | 960 | 0.008491 | 3222.8 | 3604.9 | 6.250 |
| 45 | 965 | 0.008570 | 3235.4 | 3621.0 | 6.266 |
| 45 | 970 | 0.008649 | 3247.8 | 3637.0 | 6.283 |
| 45 | 975 | 0.008727 | 3260.2 | 3652.9 | 6.299 |
| 45 | 980 | 0.008804 | 3272.5 | 3668.6 | 6.315 |
| 45 | 985 | 0.008881 | 3284.7 | 3684.3 | 6.331 |
| 45 | 990 | 0.008957 | 3296.8 | 3699.9 | 6.347 |
| 45 | 995 | 0.009033 | 3308.9 | 3715.4 | 6.363 |
| 45 | 1000 | 0.009108 | 3321.0 | 3730.9 | 6.378 |
| 45 | 1005 | 0.009183 | 3333.0 | 3746.2 | 6.394 |
| 45 | 1010 | 0.009257 | 3344.9 | 3761.5 | 6.409 |
| 45 | 1015 | 0.009331 | 3356.8 | 3776.7 | 6.424 |
| 45 | 1020 | 0.009405 | 3368.6 | 3791.8 | 6.439 |

Thermodynamic Properties of Supercritical Steam

| p | t | v | u | h | s |
|---|---|---|---|---|---|
| 45 | 1025 | 0.009478 | 3380.4 | 3806.9 | 6.453 |
| 45 | 1030 | 0.009550 | 3392.2 | 3821.9 | 6.468 |
| 45 | 1035 | 0.009622 | 3403.9 | 3836.9 | 6.482 |
| 45 | 1040 | 0.009694 | 3415.5 | 3851.8 | 6.497 |
| 45 | 1045 | 0.009765 | 3427.2 | 3866.6 | 6.511 |
| 45 | 1050 | 0.009836 | 3438.8 | 3881.4 | 6.525 |
| 45 | 1055 | 0.009907 | 3450.3 | 3896.2 | 6.539 |
| 45 | 1060 | 0.009977 | 3461.9 | 3910.8 | 6.553 |
| 45 | 1065 | 0.01005 | 3473.4 | 3925.5 | 6.567 |
| 45 | 1070 | 0.01012 | 3484.9 | 3940.1 | 6.581 |

Pressure, p = 50 Mpa

| 50 | 275 | 0.0009769 | 7.728 | 56.57 | 0.02614 |
|---|---|---|---|---|---|
| 50 | 280 | 0.0009776 | 27.85 | 76.73 | 0.09879 |
| 50 | 285 | 0.0009785 | 48.01 | 96.94 | 0.1703 |
| 50 | 290 | 0.0009797 | 68.20 | 117.2 | 0.2407 |
| 50 | 295 | 0.0009810 | 88.41 | 137.5 | 0.3101 |
| 50 | 300 | 0.0009825 | 108.6 | 157.8 | 0.3783 |
| 50 | 305 | 0.0009841 | 128.9 | 178.1 | 0.4455 |

| p | t | v | u | h | s |
|---|---|---|---|---|---|
| 50 | 310 | 0.0009859 | 149.1 | 198.4 | 0.5117 |
| 50 | 315 | 0.0009879 | 169.4 | 218.8 | 0.5768 |
| 50 | 320 | 0.0009900 | 189.7 | 239.2 | 0.6410 |
| 50 | 325 | 0.0009922 | 210.0 | 259.6 | 0.7043 |
| 50 | 330 | 0.0009946 | 230.3 | 280.0 | 0.7666 |
| 50 | 335 | 0.0009971 | 250.6 | 300.4 | 0.8280 |
| 50 | 340 | 0.001000 | 270.9 | 320.9 | 0.8886 |
| 50 | 345 | 0.001002 | 291.2 | 341.3 | 0.9484 |
| 50 | 350 | 0.001005 | 311.5 | 361.8 | 1.007 |
| 50 | 355 | 0.001008 | 331.9 | 382.3 | 1.065 |
| 50 | 360 | 0.001011 | 352.2 | 402.8 | 1.123 |
| 50 | 365 | 0.001015 | 372.6 | 423.4 | 1.179 |
| 50 | 370 | 0.001018 | 393.0 | 443.9 | 1.235 |
| 50 | 375 | 0.001021 | 413.4 | 464.5 | 1.291 |
| 50 | 380 | 0.001025 | 433.8 | 485.1 | 1.345 |
| 50 | 385 | 0.001029 | 454.3 | 505.7 | 1.399 |
| 50 | 390 | 0.001032 | 474.8 | 526.4 | 1.452 |
| 50 | 395 | 0.001036 | 495.2 | 547.1 | 1.505 |
| 50 | 400 | 0.001040 | 515.8 | 567.8 | 1.557 |
| 50 | 405 | 0.001045 | 536.3 | 588.5 | 1.609 |

Thermodynamic Properties of Supercritical Steam

| p | t | v | u | h | s |
|---|---|---|---|---|---|
| 50 | 410 | 0.001049 | 556.8 | 609.3 | 1.660 |
| 50 | 415 | 0.001053 | 577.4 | 630.1 | 1.710 |
| 50 | 420 | 0.001058 | 598.1 | 650.9 | 1.760 |
| 50 | 425 | 0.001063 | 618.7 | 671.8 | 1.810 |
| 50 | 430 | 0.001067 | 639.4 | 692.7 | 1.859 |
| 50 | 435 | 0.001072 | 660.1 | 713.7 | 1.907 |
| 50 | 440 | 0.001077 | 680.9 | 734.7 | 1.955 |
| 50 | 445 | 0.001083 | 701.6 | 755.8 | 2.003 |
| 50 | 450 | 0.001088 | 722.5 | 776.9 | 2.050 |
| 50 | 455 | 0.001093 | 743.4 | 798.0 | 2.097 |
| 50 | 460 | 0.001099 | 764.3 | 819.2 | 2.143 |
| 50 | 465 | 0.001105 | 785.3 | 840.5 | 2.189 |
| 50 | 470 | 0.001111 | 806.3 | 861.8 | 2.235 |
| 50 | 475 | 0.001117 | 827.4 | 883.2 | 2.280 |
| 50 | 480 | 0.001124 | 848.5 | 904.7 | 2.325 |
| 50 | 485 | 0.001130 | 869.7 | 926.2 | 2.369 |
| 50 | 490 | 0.001137 | 891.0 | 947.8 | 2.414 |
| 50 | 495 | 0.001144 | 912.4 | 969.5 | 2.458 |
| 50 | 500 | 0.001151 | 933.8 | 991.3 | 2.502 |
| 50 | 505 | 0.001158 | 955.3 | 1013.2 | 2.545 |

| p | t | v | u | h | s |
|---|---|---|---|---|---|
| 50 | 510 | 0.001166 | 976.9 | 1035.2 | 2.588 |
| 50 | 515 | 0.001174 | 998.5 | 1057.2 | 2.631 |
| 50 | 520 | 0.001182 | 1020.3 | 1079.4 | 2.674 |
| 50 | 525 | 0.001190 | 1042.2 | 1101.7 | 2.717 |
| 50 | 530 | 0.001199 | 1064.2 | 1124.1 | 2.759 |
| 50 | 535 | 0.001208 | 1086.3 | 1146.6 | 2.802 |
| 50 | 540 | 0.001217 | 1108.5 | 1169.3 | 2.844 |
| 50 | 545 | 0.001227 | 1130.8 | 1192.1 | 2.886 |
| 50 | 550 | 0.001237 | 1153.3 | 1215.1 | 2.928 |
| 50 | 555 | 0.001247 | 1175.9 | 1238.3 | 2.970 |
| 50 | 560 | 0.001258 | 1198.7 | 1261.6 | 3.012 |
| 50 | 565 | 0.001269 | 1221.6 | 1285.0 | 3.053 |
| 50 | 570 | 0.001280 | 1244.7 | 1308.7 | 3.095 |
| 50 | 575 | 0.001292 | 1268.0 | 1332.6 | 3.137 |
| 50 | 580 | 0.001305 | 1291.5 | 1356.7 | 3.179 |
| 50 | 585 | 0.001318 | 1315.1 | 1381.0 | 3.220 |
| 50 | 590 | 0.001332 | 1339.0 | 1405.6 | 3.262 |
| 50 | 595 | 0.001346 | 1363.1 | 1430.4 | 3.304 |
| 50 | 600 | 0.001361 | 1387.5 | 1455.6 | 3.346 |
| 50 | 605 | 0.001377 | 1412.1 | 1481.0 | 3.388 |

Thermodynamic Properties of Supercritical Steam

| p | t | v | u | h | s |
|---|---|---|---|---|---|
| 50 | 610 | 0.001394 | 1437.0 | 1506.7 | 3.431 |
| 50 | 615 | 0.001412 | 1462.2 | 1532.7 | 3.473 |
| 50 | 620 | 0.001430 | 1487.6 | 1559.1 | 3.516 |
| 50 | 625 | 0.001450 | 1513.5 | 1586.0 | 3.559 |
| 50 | 630 | 0.001471 | 1539.7 | 1613.2 | 3.602 |
| 50 | 635 | 0.001493 | 1566.3 | 1641.0 | 3.646 |
| 50 | 640 | 0.001517 | 1593.4 | 1669.2 | 3.691 |
| 50 | 645 | 0.001543 | 1621.0 | 1698.1 | 3.736 |
| 50 | 650 | 0.001570 | 1649.1 | 1727.6 | 3.781 |
| 50 | 655 | 0.001600 | 1677.8 | 1757.8 | 3.827 |
| 50 | 660 | 0.001632 | 1707.1 | 1788.7 | 3.874 |
| 50 | 665 | 0.001667 | 1737.1 | 1820.5 | 3.922 |
| 50 | 670 | 0.001705 | 1767.9 | 1853.2 | 3.971 |
| 50 | 675 | 0.001747 | 1799.6 | 1886.9 | 4.022 |
| 50 | 680 | 0.001793 | 1832.2 | 1921.8 | 4.073 |
| 50 | 685 | 0.001844 | 1865.8 | 1958.0 | 4.126 |
| 50 | 690 | 0.001901 | 1900.6 | 1995.6 | 4.181 |
| 50 | 695 | 0.001965 | 1936.5 | 2034.7 | 4.237 |
| 50 | 700 | 0.002036 | 1973.7 | 2075.5 | 4.296 |
| 50 | 705 | 0.002116 | 2012.1 | 2117.9 | 4.356 |

| p | t | v | u | h | s |
|---|---|---|---|---|---|
| 50 | 710 | 0.002205 | 2051.7 | 2162.0 | 4.418 |
| 50 | 715 | 0.002304 | 2092.4 | 2207.6 | 4.482 |
| 50 | 720 | 0.002413 | 2133.8 | 2254.4 | 4.548 |
| 50 | 725 | 0.002533 | 2175.5 | 2302.2 | 4.614 |
| 50 | 730 | 0.002661 | 2217.2 | 2350.3 | 4.680 |
| 50 | 735 | 0.002796 | 2258.4 | 2398.2 | 4.745 |
| 50 | 740 | 0.002937 | 2298.6 | 2445.5 | 4.809 |
| 50 | 745 | 0.003081 | 2337.6 | 2491.6 | 4.872 |
| 50 | 750 | 0.003227 | 2375.1 | 2536.4 | 4.931 |
| 50 | 755 | 0.003373 | 2411.0 | 2579.7 | 4.989 |
| 50 | 760 | 0.003518 | 2445.4 | 2621.3 | 5.044 |
| 50 | 765 | 0.003661 | 2478.0 | 2661.1 | 5.096 |
| 50 | 770 | 0.003802 | 2509.1 | 2699.2 | 5.146 |
| 50 | 775 | 0.003940 | 2538.9 | 2735.9 | 5.193 |
| 50 | 780 | 0.004076 | 2567.4 | 2771.2 | 5.239 |
| 50 | 785 | 0.004209 | 2594.6 | 2805.1 | 5.282 |
| 50 | 790 | 0.004338 | 2620.7 | 2837.5 | 5.323 |
| 50 | 795 | 0.004464 | 2645.6 | 2868.8 | 5.363 |
| 50 | 800 | 0.004587 | 2669.5 | 2898.8 | 5.400 |
| 50 | 805 | 0.004706 | 2692.6 | 2927.9 | 5.436 |

| p | t | v | u | h | s |
|---|---|---|---|---|---|
| 50 | 810 | 0.004823 | 2714.8 | 2955.9 | 5.471 |
| 50 | 815 | 0.004938 | 2736.2 | 2983.1 | 5.505 |
| 50 | 820 | 0.005049 | 2757.0 | 3009.5 | 5.537 |
| 50 | 825 | 0.005159 | 2777.2 | 3035.1 | 5.568 |
| 50 | 830 | 0.005266 | 2796.7 | 3060.0 | 5.598 |
| 50 | 835 | 0.005370 | 2815.8 | 3084.3 | 5.627 |
| 50 | 840 | 0.005473 | 2834.3 | 3107.9 | 5.656 |
| 50 | 845 | 0.005574 | 2852.4 | 3131.1 | 5.683 |
| 50 | 850 | 0.005673 | 2870.0 | 3153.7 | 5.710 |
| 50 | 855 | 0.005770 | 2887.3 | 3175.8 | 5.736 |
| 50 | 860 | 0.005865 | 2904.2 | 3197.5 | 5.761 |
| 50 | 865 | 0.005959 | 2920.8 | 3218.7 | 5.786 |
| 50 | 870 | 0.006051 | 2937.1 | 3239.6 | 5.810 |
| 50 | 875 | 0.006142 | 2953.1 | 3260.2 | 5.833 |
| 50 | 880 | 0.006232 | 2968.8 | 3280.4 | 5.856 |
| 50 | 885 | 0.006320 | 2984.3 | 3300.3 | 5.879 |
| 50 | 890 | 0.006407 | 2999.5 | 3319.8 | 5.901 |
| 50 | 895 | 0.006493 | 3014.5 | 3339.2 | 5.922 |
| 50 | 900 | 0.006578 | 3029.3 | 3358.2 | 5.944 |
| 50 | 905 | 0.006662 | 3044.0 | 3377.0 | 5.965 |

| p | t | v | u | h | s |
|---|---|---|---|---|---|
| 50 | 910 | 0.006744 | 3058.4 | 3395.6 | 5.985 |
| 50 | 915 | 0.006826 | 3072.7 | 3414.0 | 6.005 |
| 50 | 920 | 0.006907 | 3086.8 | 3432.1 | 6.025 |
| 50 | 925 | 0.006987 | 3100.8 | 3450.1 | 6.044 |
| 50 | 930 | 0.007066 | 3114.6 | 3467.9 | 6.064 |
| 50 | 935 | 0.007144 | 3128.3 | 3485.5 | 6.082 |
| 50 | 940 | 0.007222 | 3141.8 | 3502.9 | 6.101 |
| 50 | 945 | 0.007299 | 3155.3 | 3520.2 | 6.119 |
| 50 | 950 | 0.007375 | 3168.6 | 3537.3 | 6.137 |
| 50 | 955 | 0.007450 | 3181.8 | 3554.3 | 6.155 |
| 50 | 960 | 0.007524 | 3194.9 | 3571.1 | 6.173 |
| 50 | 965 | 0.007598 | 3207.9 | 3587.8 | 6.190 |
| 50 | 970 | 0.007672 | 3220.8 | 3604.4 | 6.207 |
| 50 | 975 | 0.007744 | 3233.6 | 3620.8 | 6.224 |
| 50 | 980 | 0.007817 | 3246.3 | 3637.2 | 6.241 |
| 50 | 985 | 0.007888 | 3259.0 | 3653.4 | 6.257 |
| 50 | 990 | 0.007959 | 3271.6 | 3669.5 | 6.274 |
| 50 | 995 | 0.008030 | 3284.1 | 3685.6 | 6.290 |
| 50 | 1000 | 0.008100 | 3296.5 | 3701.5 | 6.306 |
| 50 | 1005 | 0.008169 | 3308.9 | 3717.3 | 6.322 |

| p | t | v | u | h | s |
|---|---|---|---|---|---|
| 50 | 1010 | 0.008238 | 3321.2 | 3733.1 | 6.337 |
| 50 | 1015 | 0.008307 | 3333.4 | 3748.8 | 6.353 |
| 50 | 1020 | 0.008375 | 3345.6 | 3764.3 | 6.368 |
| 50 | 1025 | 0.008442 | 3357.7 | 3779.9 | 6.383 |
| 50 | 1030 | 0.008510 | 3369.8 | 3795.3 | 6.398 |
| 50 | 1035 | 0.008577 | 3381.8 | 3810.7 | 6.413 |
| 50 | 1040 | 0.008643 | 3393.8 | 3826.0 | 6.428 |
| 50 | 1045 | 0.008709 | 3405.8 | 3841.2 | 6.443 |
| 50 | 1050 | 0.008775 | 3417.7 | 3856.4 | 6.457 |
| 50 | 1055 | 0.008840 | 3429.5 | 3871.5 | 6.471 |
| 50 | 1060 | 0.008905 | 3441.3 | 3886.6 | 6.486 |
| 50 | 1065 | 0.008970 | 3453.1 | 3901.6 | 6.500 |
| 50 | 1070 | 0.009034 | 3464.9 | 3916.6 | 6.514 |

Pressure, p = 55Mpa

| p | t | v | u | h | s |
|---|---|---|---|---|---|
| 55 | 275 | 0.0009748 | 7.676 | 61.29 | 0.02555 |
| 55 | 280 | 0.0009755 | 27.73 | 81.38 | 0.09796 |
| 55 | 285 | 0.0009765 | 47.82 | 101.53 | 0.1693 |
| 55 | 290 | 0.0009777 | 67.95 | 121.7 | 0.2395 |
| 55 | 295 | 0.0009790 | 88.10 | 141.9 | 0.3086 |

| p | t | v | u | h | s |
|---|---|---|---|---|---|
| 55 | 300 | 0.0009805 | 108.3 | 162.2 | 0.3767 |
| 55 | 305 | 0.0009822 | 128.5 | 182.5 | 0.4438 |
| 55 | 310 | 0.0009840 | 148.7 | 202.8 | 0.5098 |
| 55 | 315 | 0.0009860 | 168.9 | 223.1 | 0.5748 |
| 55 | 320 | 0.0009881 | 189.1 | 243.4 | 0.6389 |
| 55 | 325 | 0.0009903 | 209.3 | 263.8 | 0.7020 |
| 55 | 330 | 0.0009927 | 229.6 | 284.2 | 0.7642 |
| 55 | 335 | 0.0009952 | 249.8 | 304.5 | 0.8255 |
| 55 | 340 | 0.0009978 | 270.1 | 324.9 | 0.8859 |
| 55 | 345 | 0.001000 | 290.3 | 345.4 | 0.9456 |
| 55 | 350 | 0.001003 | 310.6 | 365.8 | 1.004 |
| 55 | 355 | 0.001006 | 330.9 | 386.3 | 1.062 |
| 55 | 360 | 0.001009 | 351.2 | 406.7 | 1.120 |
| 55 | 365 | 0.001013 | 371.5 | 427.2 | 1.176 |
| 55 | 370 | 0.001016 | 391.9 | 447.7 | 1.232 |
| 55 | 375 | 0.001019 | 412.2 | 468.3 | 1.287 |
| 55 | 380 | 0.001023 | 432.6 | 488.8 | 1.342 |
| 55 | 385 | 0.001026 | 453.0 | 509.4 | 1.395 |
| 55 | 390 | 0.001030 | 473.4 | 530.0 | 1.449 |
| 55 | 395 | 0.001034 | 493.8 | 550.7 | 1.501 |

Thermodynamic Properties of Supercritical Steam

| p | t | v | u | h | s |
|---|---|---|---|---|---|
| 55 | 400 | 0.001038 | 514.2 | 571.3 | 1.553 |
| 55 | 405 | 0.001042 | 534.7 | 592.0 | 1.605 |
| 55 | 410 | 0.001046 | 555.2 | 612.8 | 1.655 |
| 55 | 415 | 0.001051 | 575.7 | 633.5 | 1.706 |
| 55 | 420 | 0.001055 | 596.3 | 654.3 | 1.756 |
| 55 | 425 | 0.001060 | 616.8 | 675.1 | 1.805 |
| 55 | 430 | 0.001065 | 637.4 | 696.0 | 1.854 |
| 55 | 435 | 0.001069 | 658.1 | 716.9 | 1.902 |
| 55 | 440 | 0.001074 | 678.8 | 737.9 | 1.950 |
| 55 | 445 | 0.001080 | 699.5 | 758.8 | 1.997 |
| 55 | 450 | 0.001085 | 720.2 | 779.9 | 2.044 |
| 55 | 455 | 0.001090 | 741.0 | 801.0 | 2.091 |
| 55 | 460 | 0.001096 | 761.8 | 822.1 | 2.137 |
| 55 | 465 | 0.001102 | 782.7 | 843.3 | 2.183 |
| 55 | 470 | 0.001107 | 803.6 | 864.6 | 2.228 |
| 55 | 475 | 0.001114 | 824.6 | 885.9 | 2.274 |
| 55 | 480 | 0.001120 | 845.7 | 907.2 | 2.318 |
| 55 | 485 | 0.001126 | 866.8 | 928.7 | 2.363 |
| 55 | 490 | 0.001133 | 887.9 | 950.2 | 2.407 |
| 55 | 495 | 0.001139 | 909.1 | 971.8 | 2.451 |

| p | t | v | u | h | s |
|---|---|---|---|---|---|
| 55 | 500 | 0.001146 | 930.4 | 993.5 | 2.494 |
| 55 | 505 | 0.001154 | 951.8 | 1015.2 | 2.538 |
| 55 | 510 | 0.001161 | 973.2 | 1037.1 | 2.581 |
| 55 | 515 | 0.001169 | 994.8 | 1059.0 | 2.624 |
| 55 | 520 | 0.001177 | 1016.4 | 1081.1 | 2.666 |
| 55 | 525 | 0.001185 | 1038.1 | 1103.2 | 2.709 |
| 55 | 530 | 0.001193 | 1059.9 | 1125.5 | 2.751 |
| 55 | 535 | 0.001202 | 1081.8 | 1147.9 | 2.793 |
| 55 | 540 | 0.001211 | 1103.8 | 1170.4 | 2.835 |
| 55 | 545 | 0.001220 | 1125.9 | 1193.0 | 2.876 |
| 55 | 550 | 0.001230 | 1148.2 | 1215.8 | 2.918 |
| 55 | 555 | 0.001239 | 1170.6 | 1238.7 | 2.960 |
| 55 | 560 | 0.001250 | 1193.1 | 1261.8 | 3.001 |
| 55 | 565 | 0.001260 | 1215.8 | 1285.1 | 3.042 |
| 55 | 570 | 0.001271 | 1238.6 | 1308.5 | 3.084 |
| 55 | 575 | 0.001283 | 1261.6 | 1332.1 | 3.125 |
| 55 | 580 | 0.001295 | 1284.7 | 1355.9 | 3.166 |
| 55 | 585 | 0.001307 | 1308.0 | 1379.9 | 3.207 |
| 55 | 590 | 0.001320 | 1331.5 | 1404.2 | 3.248 |
| 55 | 595 | 0.001334 | 1355.3 | 1428.6 | 3.290 |

Thermodynamic Properties of Supercritical Steam

| p | t | v | u | h | s |
|---|---|---|---|---|---|
| 55 | 600 | 0.001348 | 1379.2 | 1453.3 | 3.331 |
| 55 | 605 | 0.001363 | 1403.3 | 1478.3 | 3.373 |
| 55 | 610 | 0.001379 | 1427.7 | 1503.5 | 3.414 |
| 55 | 615 | 0.001395 | 1452.3 | 1529.0 | 3.456 |
| 55 | 620 | 0.001412 | 1477.2 | 1554.9 | 3.498 |
| 55 | 625 | 0.001430 | 1502.3 | 1581.0 | 3.540 |
| 55 | 630 | 0.001450 | 1527.8 | 1607.5 | 3.582 |
| 55 | 635 | 0.001470 | 1553.6 | 1634.4 | 3.624 |
| 55 | 640 | 0.001491 | 1579.8 | 1661.8 | 3.667 |
| 55 | 645 | 0.001514 | 1606.3 | 1689.6 | 3.711 |
| 55 | 650 | 0.001539 | 1633.3 | 1718.0 | 3.754 |
| 55 | 655 | 0.001565 | 1660.8 | 1746.8 | 3.799 |
| 55 | 660 | 0.001592 | 1688.7 | 1776.3 | 3.843 |
| 55 | 665 | 0.001622 | 1717.1 | 1806.3 | 3.889 |
| 55 | 670 | 0.001655 | 1746.1 | 1837.1 | 3.935 |
| 55 | 675 | 0.001690 | 1775.7 | 1868.6 | 3.982 |
| 55 | 680 | 0.001727 | 1805.9 | 1900.9 | 4.029 |
| 55 | 685 | 0.001769 | 1836.9 | 1934.2 | 4.078 |
| 55 | 690 | 0.001813 | 1868.6 | 1968.4 | 4.128 |
| 55 | 695 | 0.001863 | 1901.2 | 2003.6 | 4.179 |

| p | t | v | u | h | s |
|---|---|---|---|---|---|
| 55 | 700 | 0.001916 | 1934.5 | 2039.9 | 4.231 |
| 55 | 705 | 0.001975 | 1968.7 | 2077.4 | 4.284 |
| 55 | 710 | 0.002040 | 2003.8 | 2116.0 | 4.339 |
| 55 | 715 | 0.002111 | 2039.6 | 2155.7 | 4.394 |
| 55 | 720 | 0.002189 | 2076.2 | 2196.6 | 4.451 |
| 55 | 725 | 0.002274 | 2113.3 | 2238.4 | 4.509 |
| 55 | 730 | 0.002365 | 2150.9 | 2281.0 | 4.568 |
| 55 | 735 | 0.002464 | 2188.5 | 2324.0 | 4.627 |
| 55 | 740 | 0.002568 | 2226.1 | 2367.3 | 4.685 |
| 55 | 745 | 0.002677 | 2263.3 | 2410.6 | 4.743 |
| 55 | 750 | 0.002791 | 2299.9 | 2453.4 | 4.801 |
| 55 | 755 | 0.002908 | 2335.7 | 2495.6 | 4.857 |
| 55 | 760 | 0.003027 | 2370.5 | 2536.9 | 4.911 |
| 55 | 765 | 0.003147 | 2404.2 | 2577.3 | 4.964 |
| 55 | 770 | 0.003268 | 2436.7 | 2616.5 | 5.015 |
| 55 | 775 | 0.003389 | 2468.1 | 2654.5 | 5.065 |
| 55 | 780 | 0.003508 | 2498.2 | 2691.1 | 5.112 |
| 55 | 785 | 0.003627 | 2527.2 | 2726.6 | 5.157 |
| 55 | 790 | 0.003744 | 2555.2 | 2761.1 | 5.201 |
| 55 | 795 | 0.003859 | 2582.2 | 2794.4 | 5.243 |

| p | t | v | u | h | s |
|---|---|---|---|---|---|
| 55 | 800 | 0.003973 | 2608.1 | 2826.6 | 5.283 |
| 55 | 805 | 0.004084 | 2633.0 | 2857.7 | 5.322 |
| 55 | 810 | 0.004193 | 2657.1 | 2887.7 | 5.359 |
| 55 | 815 | 0.004300 | 2680.4 | 2916.9 | 5.395 |
| 55 | 820 | 0.004405 | 2702.8 | 2945.1 | 5.430 |
| 55 | 825 | 0.004508 | 2724.6 | 2972.5 | 5.463 |
| 55 | 830 | 0.004608 | 2745.8 | 2999.2 | 5.495 |
| 55 | 835 | 0.004707 | 2766.3 | 3025.2 | 5.526 |
| 55 | 840 | 0.004804 | 2786.2 | 3050.5 | 5.557 |
| 55 | 845 | 0.004900 | 2805.7 | 3075.2 | 5.586 |
| 55 | 850 | 0.004993 | 2824.6 | 3099.2 | 5.614 |
| 55 | 855 | 0.005085 | 2843.1 | 3122.8 | 5.642 |
| 55 | 860 | 0.005175 | 2861.2 | 3145.8 | 5.669 |
| 55 | 865 | 0.005264 | 2878.9 | 3168.4 | 5.695 |
| 55 | 870 | 0.005351 | 2896.2 | 3190.5 | 5.721 |
| 55 | 875 | 0.005437 | 2913.2 | 3212.3 | 5.745 |
| 55 | 880 | 0.005522 | 2929.9 | 3233.6 | 5.770 |
| 55 | 885 | 0.005606 | 2946.3 | 3254.6 | 5.793 |
| 55 | 890 | 0.005688 | 2962.4 | 3275.2 | 5.817 |
| 55 | 895 | 0.005769 | 2978.2 | 3295.5 | 5.840 |

| p | t | v | u | h | s |
|---|---|---|---|---|---|
| 55 | 900 | 0.005849 | 2993.8 | 3315.5 | 5.862 |
| 55 | 905 | 0.005928 | 3009.2 | 3335.3 | 5.884 |
| 55 | 910 | 0.006006 | 3024.4 | 3354.7 | 5.905 |
| 55 | 915 | 0.006083 | 3039.4 | 3374.0 | 5.926 |
| 55 | 920 | 0.006160 | 3054.2 | 3393.0 | 5.947 |
| 55 | 925 | 0.006235 | 3068.8 | 3411.7 | 5.967 |
| 55 | 930 | 0.006309 | 3083.2 | 3430.3 | 5.987 |
| 55 | 935 | 0.006383 | 3097.5 | 3448.6 | 6.007 |
| 55 | 940 | 0.006456 | 3111.7 | 3466.8 | 6.026 |
| 55 | 945 | 0.006528 | 3125.7 | 3484.7 | 6.045 |
| 55 | 950 | 0.006600 | 3139.5 | 3502.5 | 6.064 |
| 55 | 955 | 0.006670 | 3153.3 | 3520.1 | 6.083 |
| 55 | 960 | 0.006740 | 3166.9 | 3537.6 | 6.101 |
| 55 | 965 | 0.006810 | 3180.4 | 3554.9 | 6.119 |
| 55 | 970 | 0.006879 | 3193.8 | 3572.1 | 6.137 |
| 55 | 975 | 0.006947 | 3207.0 | 3589.1 | 6.154 |
| 55 | 980 | 0.007015 | 3220.2 | 3606.0 | 6.171 |
| 55 | 985 | 0.007082 | 3233.3 | 3622.8 | 6.188 |
| 55 | 990 | 0.007148 | 3246.3 | 3639.5 | 6.205 |
| 55 | 995 | 0.007214 | 3259.2 | 3656.0 | 6.222 |

| p | t | v | u | h | s |
|---|---|---|---|---|---|
| 55 | 1000 | 0.007280 | 3272.0 | 3672.4 | 6.238 |
| 55 | 1005 | 0.007345 | 3284.8 | 3688.7 | 6.255 |
| 55 | 1010 | 0.007409 | 3297.5 | 3705.0 | 6.271 |
| 55 | 1015 | 0.007474 | 3310.1 | 3721.1 | 6.287 |
| 55 | 1020 | 0.007537 | 3322.6 | 3737.1 | 6.303 |
| 55 | 1025 | 0.007600 | 3335.1 | 3753.1 | 6.318 |
| 55 | 1030 | 0.007663 | 3347.5 | 3769.0 | 6.334 |
| 55 | 1035 | 0.007726 | 3359.8 | 3784.7 | 6.349 |
| 55 | 1040 | 0.007788 | 3372.1 | 3800.5 | 6.364 |
| 55 | 1045 | 0.007849 | 3384.4 | 3816.1 | 6.379 |
| 55 | 1050 | 0.007911 | 3396.6 | 3831.7 | 6.394 |
| 55 | 1055 | 0.007972 | 3408.7 | 3847.2 | 6.409 |
| 55 | 1060 | 0.008032 | 3420.8 | 3862.6 | 6.423 |
| 55 | 1065 | 0.008092 | 3432.9 | 3878.0 | 6.438 |
| 55 | 1070 | 0.008152 | 3444.9 | 3893.3 | 6.452 |

Pressure, p = 60Mpa

| 60 | 275 | 0.0009727 | 7.617 | 65.98 | 0.02490 |
|---|---|---|---|---|---|
| 60 | 280 | 0.0009735 | 27.60 | 86.01 | 0.09708 |

| p | t | v | u | h | s |
|---|---|---|---|---|---|
| 60 | 285 | 0.0009745 | 47.63 | 106.10 | 0.1682 |
| 60 | 290 | 0.0009757 | 67.69 | 126.2 | 0.2382 |
| 60 | 295 | 0.0009771 | 87.78 | 146.4 | 0.3072 |
| 60 | 300 | 0.0009786 | 107.9 | 166.6 | 0.3751 |
| 60 | 305 | 0.0009803 | 128.0 | 186.8 | 0.4420 |
| 60 | 310 | 0.0009821 | 148.2 | 207.1 | 0.5079 |
| 60 | 315 | 0.0009841 | 168.3 | 227.4 | 0.5728 |
| 60 | 320 | 0.0009862 | 188.5 | 247.7 | 0.6367 |
| 60 | 325 | 0.0009884 | 208.7 | 268.0 | 0.6997 |
| 60 | 330 | 0.0009908 | 228.9 | 288.3 | 0.7618 |
| 60 | 335 | 0.0009932 | 249.1 | 308.7 | 0.8230 |
| 60 | 340 | 0.0009958 | 269.3 | 329.0 | 0.8833 |
| 60 | 345 | 0.000999 | 289.5 | 349.4 | 0.9428 |
| 60 | 350 | 0.001001 | 309.7 | 369.8 | 1.001 |
| 60 | 355 | 0.001004 | 330.0 | 390.2 | 1.059 |
| 60 | 360 | 0.001007 | 350.2 | 410.7 | 1.117 |
| 60 | 365 | 0.001011 | 370.5 | 431.1 | 1.173 |
| 60 | 370 | 0.001014 | 390.8 | 451.6 | 1.229 |
| 60 | 375 | 0.001017 | 411.0 | 472.1 | 1.284 |
| 60 | 380 | 0.001021 | 431.4 | 492.6 | 1.338 |

Thermodynamic Properties of Supercritical Steam

| p | t | v | u | h | s |
|---|---|---|---|---|---|
| 60 | 385 | 0.001024 | 451.7 | 513.1 | 1.392 |
| 60 | 390 | 0.001028 | 472.0 | 533.7 | 1.445 |
| 60 | 395 | 0.001032 | 492.4 | 554.3 | 1.497 |
| 60 | 400 | 0.001036 | 512.8 | 574.9 | 1.549 |
| 60 | 405 | 0.001040 | 533.2 | 595.6 | 1.600 |
| 60 | 410 | 0.001044 | 553.6 | 616.2 | 1.651 |
| 60 | 415 | 0.001048 | 574.0 | 636.9 | 1.701 |
| 60 | 420 | 0.001053 | 594.5 | 657.7 | 1.751 |
| 60 | 425 | 0.001057 | 615.0 | 678.5 | 1.800 |
| 60 | 430 | 0.001062 | 635.6 | 699.3 | 1.849 |
| 60 | 435 | 0.001067 | 656.1 | 720.1 | 1.897 |
| 60 | 440 | 0.001072 | 676.7 | 741.0 | 1.945 |
| 60 | 445 | 0.001077 | 697.3 | 761.9 | 1.992 |
| 60 | 450 | 0.001082 | 718.0 | 782.9 | 2.039 |
| 60 | 455 | 0.001087 | 738.7 | 803.9 | 2.086 |
| 60 | 460 | 0.001093 | 759.4 | 825.0 | 2.132 |
| 60 | 465 | 0.001098 | 780.2 | 846.1 | 2.177 |
| 60 | 470 | 0.001104 | 801.1 | 867.3 | 2.223 |
| 60 | 475 | 0.001110 | 821.9 | 888.5 | 2.268 |
| 60 | 480 | 0.001116 | 842.9 | 909.8 | 2.312 |

| p | t | v | u | h | s |
|---|---|---|---|---|---|
| 60 | 485 | 0.001122 | 863.9 | 931.2 | 2.356 |
| 60 | 490 | 0.001129 | 884.9 | 952.6 | 2.400 |
| 60 | 495 | 0.001135 | 906.0 | 974.1 | 2.444 |
| 60 | 500 | 0.001142 | 927.2 | 995.7 | 2.487 |
| 60 | 505 | 0.001149 | 948.4 | 1017.4 | 2.530 |
| 60 | 510 | 0.001156 | 969.7 | 1039.1 | 2.573 |
| 60 | 515 | 0.001164 | 991.1 | 1060.9 | 2.616 |
| 60 | 520 | 0.001171 | 1012.6 | 1082.9 | 2.658 |
| 60 | 525 | 0.001179 | 1034.1 | 1104.9 | 2.700 |
| 60 | 530 | 0.001188 | 1055.8 | 1127.0 | 2.742 |
| 60 | 535 | 0.001196 | 1077.5 | 1149.2 | 2.784 |
| 60 | 540 | 0.001205 | 1099.3 | 1171.6 | 2.826 |
| 60 | 545 | 0.001214 | 1121.3 | 1194.1 | 2.867 |
| 60 | 550 | 0.001223 | 1143.3 | 1216.7 | 2.908 |
| 60 | 555 | 0.001232 | 1165.5 | 1239.4 | 2.950 |
| 60 | 560 | 0.001242 | 1187.8 | 1262.3 | 2.991 |
| 60 | 565 | 0.001253 | 1210.2 | 1285.4 | 3.032 |
| 60 | 570 | 0.001263 | 1232.8 | 1308.6 | 3.072 |
| 60 | 575 | 0.001274 | 1255.5 | 1331.9 | 3.113 |
| 60 | 580 | 0.001286 | 1278.3 | 1355.5 | 3.154 |

Thermodynamic Properties of Supercritical Steam

| p | t | v | u | h | s |
|---|---|---|---|---|---|
| 60 | 585 | 0.001297 | 1301.3 | 1379.2 | 3.195 |
| 60 | 590 | 0.001310 | 1324.5 | 1403.1 | 3.236 |
| 60 | 595 | 0.001323 | 1347.9 | 1427.2 | 3.276 |
| 60 | 600 | 0.001336 | 1371.4 | 1451.6 | 3.317 |
| 60 | 605 | 0.001350 | 1395.1 | 1476.1 | 3.358 |
| 60 | 610 | 0.001365 | 1419.1 | 1501.0 | 3.399 |
| 60 | 615 | 0.001380 | 1443.2 | 1526.0 | 3.440 |
| 60 | 620 | 0.001396 | 1467.6 | 1551.3 | 3.481 |
| 60 | 625 | 0.001413 | 1492.1 | 1576.9 | 3.522 |
| 60 | 630 | 0.001431 | 1517.0 | 1602.8 | 3.563 |
| 60 | 635 | 0.001449 | 1542.1 | 1629.1 | 3.604 |
| 60 | 640 | 0.001469 | 1567.5 | 1655.7 | 3.646 |
| 60 | 645 | 0.001490 | 1593.3 | 1682.7 | 3.688 |
| 60 | 650 | 0.001512 | 1619.4 | 1710.1 | 3.730 |
| 60 | 655 | 0.001535 | 1645.8 | 1737.9 | 3.773 |
| 60 | 660 | 0.001560 | 1672.7 | 1766.2 | 3.816 |
| 60 | 665 | 0.001586 | 1699.9 | 1795.0 | 3.860 |
| 60 | 670 | 0.001614 | 1727.5 | 1824.4 | 3.904 |
| 60 | 675 | 0.001644 | 1755.6 | 1854.3 | 3.948 |
| 60 | 680 | 0.001677 | 1784.2 | 1884.8 | 3.993 |

| p | t | v | u | h | s |
|---|---|---|---|---|---|
| 60 | 685 | 0.001711 | 1813.3 | 1916.0 | 4.039 |
| 60 | 690 | 0.001749 | 1843.0 | 1947.9 | 4.085 |
| 60 | 695 | 0.001789 | 1873.3 | 1980.6 | 4.132 |
| 60 | 700 | 0.001832 | 1904.1 | 2014.0 | 4.180 |
| 60 | 705 | 0.001879 | 1935.6 | 2048.3 | 4.229 |
| 60 | 710 | 0.001930 | 1967.6 | 2083.4 | 4.279 |
| 60 | 715 | 0.001985 | 2000.3 | 2119.4 | 4.329 |
| 60 | 720 | 0.002045 | 2033.5 | 2156.2 | 4.381 |
| 60 | 725 | 0.002109 | 2067.3 | 2193.8 | 4.433 |
| 60 | 730 | 0.002178 | 2101.4 | 2232.1 | 4.485 |
| 60 | 735 | 0.002252 | 2135.9 | 2271.1 | 4.538 |
| 60 | 740 | 0.002331 | 2170.6 | 2310.5 | 4.592 |
| 60 | 745 | 0.002415 | 2205.2 | 2350.1 | 4.645 |
| 60 | 750 | 0.002503 | 2239.8 | 2390.0 | 4.699 |
| 60 | 755 | 0.002595 | 2274.0 | 2429.7 | 4.751 |
| 60 | 760 | 0.002690 | 2307.7 | 2469.1 | 4.803 |
| 60 | 765 | 0.002788 | 2340.9 | 2508.1 | 4.855 |
| 60 | 770 | 0.002888 | 2373.3 | 2546.6 | 4.905 |
| 60 | 775 | 0.002989 | 2404.9 | 2584.3 | 4.954 |
| 60 | 780 | 0.003091 | 2435.7 | 2621.2 | 5.001 |

Thermodynamic Properties of Supercritical Steam

| p | t | v | u | h | s |
|---|---|---|---|---|---|
| 60 | 785 | 0.003194 | 2465.6 | 2657.2 | 5.047 |
| 60 | 790 | 0.003297 | 2494.5 | 2692.3 | 5.092 |
| 60 | 795 | 0.003398 | 2522.5 | 2726.4 | 5.135 |
| 60 | 800 | 0.003500 | 2549.7 | 2759.7 | 5.176 |
| 60 | 805 | 0.003600 | 2576.1 | 2792.1 | 5.217 |
| 60 | 810 | 0.003700 | 2601.6 | 2823.6 | 5.256 |
| 60 | 815 | 0.003798 | 2626.2 | 2854.1 | 5.293 |
| 60 | 820 | 0.003895 | 2650.1 | 2883.8 | 5.330 |
| 60 | 825 | 0.003990 | 2673.3 | 2912.7 | 5.365 |
| 60 | 830 | 0.004084 | 2695.7 | 2940.8 | 5.399 |
| 60 | 835 | 0.004176 | 2717.6 | 2968.1 | 5.432 |
| 60 | 840 | 0.004267 | 2738.8 | 2994.8 | 5.463 |
| 60 | 845 | 0.004356 | 2759.5 | 3020.9 | 5.494 |
| 60 | 850 | 0.004444 | 2779.6 | 3046.3 | 5.524 |
| 60 | 855 | 0.004531 | 2799.3 | 3071.1 | 5.554 |
| 60 | 860 | 0.004616 | 2818.5 | 3095.4 | 5.582 |
| 60 | 865 | 0.004700 | 2837.2 | 3119.2 | 5.609 |
| 60 | 870 | 0.004782 | 2855.6 | 3142.5 | 5.636 |
| 60 | 875 | 0.004863 | 2873.5 | 3165.4 | 5.662 |
| 60 | 880 | 0.004944 | 2891.2 | 3187.8 | 5.688 |

| p | t | v | u | h | s |
|---|---|---|---|---|---|
| 60 | 885 | 0.005022 | 2908.4 | 3209.8 | 5.713 |
| 60 | 890 | 0.005100 | 2925.4 | 3231.4 | 5.737 |
| 60 | 895 | 0.005177 | 2942.1 | 3252.7 | 5.761 |
| 60 | 900 | 0.005253 | 2958.5 | 3273.6 | 5.784 |
| 60 | 905 | 0.005327 | 2974.6 | 3294.3 | 5.807 |
| 60 | 910 | 0.005401 | 2990.5 | 3314.6 | 5.830 |
| 60 | 915 | 0.005474 | 3006.2 | 3334.6 | 5.852 |
| 60 | 920 | 0.005546 | 3021.7 | 3354.4 | 5.873 |
| 60 | 925 | 0.005617 | 3036.9 | 3374.0 | 5.894 |
| 60 | 930 | 0.005687 | 3052.0 | 3393.3 | 5.915 |
| 60 | 935 | 0.005757 | 3066.9 | 3412.3 | 5.936 |
| 60 | 940 | 0.005826 | 3081.6 | 3431.2 | 5.956 |
| 60 | 945 | 0.005894 | 3096.2 | 3449.8 | 5.976 |
| 60 | 950 | 0.005961 | 3110.6 | 3468.3 | 5.995 |
| 60 | 955 | 0.006028 | 3124.9 | 3486.5 | 6.014 |
| 60 | 960 | 0.006094 | 3139.0 | 3504.6 | 6.033 |
| 60 | 965 | 0.006160 | 3153.0 | 3522.5 | 6.052 |
| 60 | 970 | 0.006225 | 3166.8 | 3540.3 | 6.070 |
| 60 | 975 | 0.006289 | 3180.6 | 3557.9 | 6.088 |
| 60 | 980 | 0.006353 | 3194.2 | 3575.3 | 6.106 |

Thermodynamic Properties of Supercritical Steam

| p | t | v | u | h | s |
|---|---|---|---|---|---|
| 60 | 985 | 0.006416 | 3207.7 | 3592.6 | 6.124 |
| 60 | 990 | 0.006478 | 3221.1 | 3609.8 | 6.141 |
| 60 | 995 | 0.006541 | 3234.4 | 3626.8 | 6.158 |
| 60 | 1000 | 0.006602 | 3247.6 | 3643.8 | 6.175 |
| 60 | 1005 | 0.006663 | 3260.8 | 3660.6 | 6.192 |
| 60 | 1010 | 0.006724 | 3273.8 | 3677.3 | 6.208 |
| 60 | 1015 | 0.006784 | 3286.8 | 3693.8 | 6.225 |
| 60 | 1020 | 0.006844 | 3299.6 | 3710.3 | 6.241 |
| 60 | 1025 | 0.006904 | 3312.5 | 3726.7 | 6.257 |
| 60 | 1030 | 0.006963 | 3325.2 | 3743.0 | 6.273 |
| 60 | 1035 | 0.007021 | 3337.9 | 3759.2 | 6.289 |
| 60 | 1040 | 0.007080 | 3350.5 | 3775.3 | 6.304 |
| 60 | 1045 | 0.007138 | 3363.0 | 3791.3 | 6.319 |
| 60 | 1050 | 0.007195 | 3375.5 | 3807.2 | 6.335 |
| 60 | 1055 | 0.007252 | 3388.0 | 3823.1 | 6.350 |
| 60 | 1060 | 0.007309 | 3400.4 | 3838.9 | 6.365 |
| 60 | 1065 | 0.007365 | 3412.7 | 3854.6 | 6.380 |
| 60 | 1070 | 0.007422 | 3425.0 | 3870.3 | 6.394 |

| p | t | v | u | h | s |
|---|---|---|---|---|---|

Pressure, p = 65 Mpa

| p | t | v | u | h | s |
|---|---|---|---|---|---|
| 65 | 275 | 0.0009706 | 7.551 | 70.64 | 0.02419 |
| 65 | 280 | 0.0009715 | 27.47 | 90.61 | 0.09615 |
| 65 | 285 | 0.0009725 | 47.43 | 110.6 | 0.1671 |
| 65 | 290 | 0.0009738 | 67.44 | 130.7 | 0.2369 |
| 65 | 295 | 0.0009752 | 87.47 | 150.9 | 0.3057 |
| 65 | 300 | 0.0009767 | 107.5 | 171.0 | 0.3735 |
| 65 | 305 | 0.0009784 | 127.6 | 191.2 | 0.4403 |
| 65 | 310 | 0.0009802 | 147.7 | 211.4 | 0.5060 |
| 65 | 315 | 0.0009822 | 167.8 | 231.7 | 0.5707 |
| 65 | 320 | 0.0009843 | 187.9 | 251.9 | 0.6345 |
| 65 | 325 | 0.0009865 | 208.1 | 272.2 | 0.6974 |
| 65 | 330 | 0.0009889 | 228.2 | 292.5 | 0.7594 |
| 65 | 335 | 0.0009914 | 248.3 | 312.8 | 0.8204 |
| 65 | 340 | 0.0009939 | 268.5 | 333.1 | 0.8806 |
| 65 | 345 | 0.0009966 | 288.7 | 353.4 | 0.9400 |
| 65 | 350 | 0.0009994 | 308.8 | 373.8 | 0.9986 |
| 65 | 355 | 0.001002 | 329.0 | 394.2 | 1.056 |
| 65 | 360 | 0.001005 | 349.2 | 414.6 | 1.113 |

| p | t | v | u | h | s |
|---|---|---|---|---|---|
| 65 | 365 | 0.001009 | 369.4 | 435.0 | 1.170 |
| 65 | 370 | 0.001012 | 389.7 | 455.4 | 1.225 |
| 65 | 375 | 0.001015 | 409.9 | 475.9 | 1.280 |
| 65 | 380 | 0.001019 | 430.1 | 496.4 | 1.335 |
| 65 | 385 | 0.001022 | 450.4 | 516.9 | 1.388 |
| 65 | 390 | 0.001026 | 470.7 | 537.4 | 1.441 |
| 65 | 395 | 0.001030 | 491.0 | 557.9 | 1.493 |
| 65 | 400 | 0.001033 | 511.3 | 578.5 | 1.545 |
| 65 | 405 | 0.001037 | 531.7 | 599.1 | 1.596 |
| 65 | 410 | 0.001042 | 552.0 | 619.7 | 1.647 |
| 65 | 415 | 0.001046 | 572.4 | 640.4 | 1.697 |
| 65 | 420 | 0.001050 | 592.8 | 661.1 | 1.747 |
| 65 | 425 | 0.001055 | 613.2 | 681.8 | 1.796 |
| 65 | 430 | 0.001059 | 633.7 | 702.6 | 1.844 |
| 65 | 435 | 0.001064 | 654.2 | 723.4 | 1.892 |
| 65 | 440 | 0.001069 | 674.7 | 744.2 | 1.940 |
| 65 | 445 | 0.001074 | 695.3 | 765.1 | 1.987 |
| 65 | 450 | 0.001079 | 715.8 | 786.0 | 2.034 |
| 65 | 455 | 0.001084 | 736.5 | 806.9 | 2.080 |
| 65 | 460 | 0.001089 | 757.1 | 827.9 | 2.126 |

| p | t | v | u | h | s |
|---|---|---|---|---|---|
| 65 | 465 | 0.001095 | 777.8 | 849.0 | 2.172 |
| 65 | 470 | 0.001101 | 798.5 | 870.1 | 2.217 |
| 65 | 475 | 0.001106 | 819.3 | 891.3 | 2.262 |
| 65 | 480 | 0.001112 | 840.2 | 912.5 | 2.306 |
| 65 | 485 | 0.001119 | 861.0 | 933.7 | 2.350 |
| 65 | 490 | 0.001125 | 882.0 | 955.1 | 2.394 |
| 65 | 495 | 0.001131 | 903.0 | 976.5 | 2.437 |
| 65 | 500 | 0.001138 | 924.0 | 998.0 | 2.481 |
| 65 | 505 | 0.001145 | 945.1 | 1019.5 | 2.523 |
| 65 | 510 | 0.001152 | 966.3 | 1041.2 | 2.566 |
| 65 | 515 | 0.001159 | 987.6 | 1062.9 | 2.608 |
| 65 | 520 | 0.001167 | 1008.9 | 1084.7 | 2.651 |
| 65 | 525 | 0.001174 | 1030.3 | 1106.6 | 2.693 |
| 65 | 530 | 0.001182 | 1051.8 | 1128.6 | 2.734 |
| 65 | 535 | 0.001190 | 1073.3 | 1150.7 | 2.776 |
| 65 | 540 | 0.001199 | 1095.0 | 1172.9 | 2.817 |
| 65 | 545 | 0.001207 | 1116.8 | 1195.3 | 2.858 |
| 65 | 550 | 0.001216 | 1138.6 | 1217.7 | 2.899 |
| 65 | 555 | 0.001226 | 1160.6 | 1240.3 | 2.940 |
| 65 | 560 | 0.001235 | 1182.7 | 1263.0 | 2.981 |

| p | t | v | u | h | s |
|---|---|---|---|---|---|
| 65 | 565 | 0.001245 | 1204.9 | 1285.8 | 3.021 |
| 65 | 570 | 0.001255 | 1227.2 | 1308.8 | 3.062 |
| 65 | 575 | 0.001266 | 1249.7 | 1331.9 | 3.102 |
| 65 | 580 | 0.001277 | 1272.3 | 1355.2 | 3.143 |
| 65 | 585 | 0.001288 | 1295.0 | 1378.7 | 3.183 |
| 65 | 590 | 0.001300 | 1317.9 | 1402.4 | 3.223 |
| 65 | 595 | 0.001312 | 1340.9 | 1426.2 | 3.263 |
| 65 | 600 | 0.001325 | 1364.1 | 1450.2 | 3.304 |
| 65 | 605 | 0.001338 | 1387.5 | 1474.4 | 3.344 |
| 65 | 610 | 0.001352 | 1411.0 | 1498.9 | 3.384 |
| 65 | 615 | 0.001366 | 1434.7 | 1523.5 | 3.424 |
| 65 | 620 | 0.001382 | 1458.6 | 1548.4 | 3.465 |
| 65 | 625 | 0.001397 | 1482.7 | 1573.5 | 3.505 |
| 65 | 630 | 0.001414 | 1507.0 | 1598.9 | 3.545 |
| 65 | 635 | 0.001431 | 1531.6 | 1624.6 | 3.586 |
| 65 | 640 | 0.001449 | 1556.4 | 1650.6 | 3.627 |
| 65 | 645 | 0.001468 | 1581.5 | 1676.9 | 3.668 |
| 65 | 650 | 0.001488 | 1606.8 | 1703.6 | 3.709 |
| 65 | 655 | 0.001509 | 1632.5 | 1730.6 | 3.750 |
| 65 | 660 | 0.001532 | 1658.4 | 1758.0 | 3.792 |

| p | t | v | u | h | s |
|---|---|---|---|---|---|
| 65 | 665 | 0.001555 | 1684.7 | 1785.8 | 3.834 |
| 65 | 670 | 0.001580 | 1711.3 | 1814.0 | 3.876 |
| 65 | 675 | 0.001607 | 1738.3 | 1842.7 | 3.919 |
| 65 | 680 | 0.001635 | 1765.6 | 1871.9 | 3.962 |
| 65 | 685 | 0.001665 | 1793.4 | 1901.6 | 4.006 |
| 65 | 690 | 0.001697 | 1821.5 | 1931.9 | 4.050 |
| 65 | 695 | 0.001732 | 1850.1 | 1962.7 | 4.094 |
| 65 | 700 | 0.001768 | 1879.2 | 1994.1 | 4.139 |
| 65 | 705 | 0.001807 | 1908.7 | 2026.2 | 4.185 |
| 65 | 710 | 0.001849 | 1938.7 | 2058.9 | 4.231 |
| 65 | 715 | 0.001894 | 1969.1 | 2092.3 | 4.278 |
| 65 | 720 | 0.001943 | 2000.0 | 2126.3 | 4.325 |
| 65 | 725 | 0.001994 | 2031.3 | 2160.9 | 4.373 |
| 65 | 730 | 0.002049 | 2063.0 | 2196.2 | 4.422 |
| 65 | 735 | 0.002108 | 2094.9 | 2231.9 | 4.470 |
| 65 | 740 | 0.002170 | 2127.1 | 2268.2 | 4.520 |
| 65 | 745 | 0.002236 | 2159.5 | 2304.9 | 4.569 |
| 65 | 750 | 0.002306 | 2191.9 | 2341.8 | 4.618 |
| 65 | 755 | 0.002379 | 2224.2 | 2378.9 | 4.668 |
| 65 | 760 | 0.002456 | 2256.4 | 2416.0 | 4.717 |

Thermodynamic Properties of Supercritical Steam

| p | t | v | u | h | s |
|---|---|---|---|---|---|
| 65 | 765 | 0.002535 | 2288.2 | 2453.0 | 4.765 |
| 65 | 770 | 0.002617 | 2319.7 | 2489.8 | 4.813 |
| 65 | 775 | 0.002701 | 2350.7 | 2526.3 | 4.860 |
| 65 | 780 | 0.002786 | 2381.2 | 2562.3 | 4.907 |
| 65 | 785 | 0.002873 | 2411.0 | 2597.7 | 4.952 |
| 65 | 790 | 0.002961 | 2440.1 | 2632.6 | 4.996 |
| 65 | 795 | 0.003050 | 2468.6 | 2666.8 | 5.039 |
| 65 | 800 | 0.003139 | 2496.4 | 2700.4 | 5.082 |
| 65 | 805 | 0.003227 | 2523.3 | 2733.1 | 5.122 |
| 65 | 810 | 0.003316 | 2549.6 | 2765.2 | 5.162 |
| 65 | 815 | 0.003404 | 2575.2 | 2796.5 | 5.201 |
| 65 | 820 | 0.003492 | 2600.1 | 2827.0 | 5.238 |
| 65 | 825 | 0.003579 | 2624.2 | 2856.9 | 5.274 |
| 65 | 830 | 0.003665 | 2647.7 | 2885.9 | 5.309 |
| 65 | 835 | 0.003750 | 2670.6 | 2914.3 | 5.343 |
| 65 | 840 | 0.003834 | 2692.9 | 2942.1 | 5.377 |
| 65 | 845 | 0.003917 | 2714.6 | 2969.2 | 5.409 |
| 65 | 850 | 0.003998 | 2735.8 | 2995.7 | 5.440 |
| 65 | 855 | 0.004079 | 2756.5 | 3021.6 | 5.470 |
| 65 | 860 | 0.004159 | 2776.6 | 3047.0 | 5.500 |

| p | t | v | u | h | s |
|---|---|---|---|---|---|
| 65 | 865 | 0.004238 | 2796.4 | 3071.8 | 5.529 |
| 65 | 870 | 0.004315 | 2815.6 | 3096.1 | 5.557 |
| 65 | 875 | 0.004392 | 2834.5 | 3120.0 | 5.584 |
| 65 | 880 | 0.004467 | 2853.0 | 3143.3 | 5.611 |
| 65 | 885 | 0.004541 | 2871.1 | 3166.3 | 5.637 |
| 65 | 890 | 0.004615 | 2888.9 | 3188.8 | 5.662 |
| 65 | 895 | 0.004687 | 2906.3 | 3211.0 | 5.687 |
| 65 | 900 | 0.004759 | 2923.5 | 3232.8 | 5.711 |
| 65 | 905 | 0.004829 | 2940.4 | 3254.3 | 5.735 |
| 65 | 910 | 0.004899 | 2957.0 | 3275.4 | 5.758 |
| 65 | 915 | 0.004968 | 2973.4 | 3296.3 | 5.781 |
| 65 | 920 | 0.005036 | 2989.5 | 3316.8 | 5.804 |
| 65 | 925 | 0.005103 | 3005.4 | 3337.1 | 5.826 |
| 65 | 930 | 0.005170 | 3021.1 | 3357.1 | 5.847 |
| 65 | 935 | 0.005236 | 3036.5 | 3376.8 | 5.868 |
| 65 | 940 | 0.005301 | 3051.8 | 3396.4 | 5.889 |
| 65 | 945 | 0.005365 | 3066.9 | 3415.7 | 5.910 |
| 65 | 950 | 0.005429 | 3081.9 | 3434.8 | 5.930 |
| 65 | 955 | 0.005492 | 3096.7 | 3453.6 | 5.950 |
| 65 | 960 | 0.005554 | 3111.3 | 3472.3 | 5.969 |

Thermodynamic Properties of Supercritical Steam

| p | t | v | u | h | s |
|---|---|---|---|---|---|
| 65 | 965 | 0.005616 | 3125.7 | 3490.8 | 5.988 |
| 65 | 970 | 0.005678 | 3140.1 | 3509.1 | 6.007 |
| 65 | 975 | 0.005738 | 3154.2 | 3527.2 | 6.026 |
| 65 | 980 | 0.005799 | 3168.3 | 3545.2 | 6.044 |
| 65 | 985 | 0.005858 | 3182.2 | 3563.0 | 6.062 |
| 65 | 990 | 0.005918 | 3196.1 | 3580.7 | 6.080 |
| 65 | 995 | 0.005976 | 3209.8 | 3598.2 | 6.098 |
| 65 | 1000 | 0.006035 | 3223.4 | 3615.6 | 6.115 |
| 65 | 1005 | 0.006092 | 3236.9 | 3632.9 | 6.133 |
| 65 | 1010 | 0.006150 | 3250.3 | 3650.0 | 6.150 |
| 65 | 1015 | 0.006207 | 3263.6 | 3667.0 | 6.166 |
| 65 | 1020 | 0.006263 | 3276.8 | 3683.9 | 6.183 |
| 65 | 1025 | 0.006319 | 3290.0 | 3700.7 | 6.200 |
| 65 | 1030 | 0.006375 | 3303.0 | 3717.4 | 6.216 |
| 65 | 1035 | 0.006430 | 3316.0 | 3734.0 | 6.232 |
| 65 | 1040 | 0.006485 | 3329.0 | 3750.5 | 6.248 |
| 65 | 1045 | 0.006540 | 3341.8 | 3766.9 | 6.263 |
| 65 | 1050 | 0.006594 | 3354.6 | 3783.2 | 6.279 |
| 65 | 1055 | 0.006648 | 3367.4 | 3799.5 | 6.294 |
| 65 | 1060 | 0.006701 | 3380.0 | 3815.6 | 6.310 |

| p | t | v | u | h | s |
|---|---|---|---|---|---|
| 65 | 1065 | 0.006755 | 3392.7 | 3831.7 | 6.325 |
| 65 | 1070 | 0.006808 | 3405.2 | 3847.7 | 6.340 |

## Pressure, p = 70 Mpa

| | | | | | |
|---|---|---|---|---|---|
| 70 | 275 | 0.0009686 | 7.478 | 75.28 | 0.02343 |
| 70 | 280 | 0.0009695 | 27.33 | 95.19 | 0.09518 |
| 70 | 285 | 0.0009706 | 47.23 | 115.2 | 0.1659 |
| 70 | 290 | 0.0009719 | 67.18 | 135.2 | 0.2356 |
| 70 | 295 | 0.0009733 | 87.16 | 155.3 | 0.3043 |
| 70 | 300 | 0.0009748 | 107.2 | 175.4 | 0.3719 |
| 70 | 305 | 0.0009766 | 127.2 | 195.6 | 0.4385 |
| 70 | 310 | 0.0009784 | 147.2 | 215.7 | 0.5041 |
| 70 | 315 | 0.0009804 | 167.3 | 235.9 | 0.5687 |
| 70 | 320 | 0.0009825 | 187.4 | 256.1 | 0.6324 |
| 70 | 325 | 0.0009847 | 207.4 | 276.4 | 0.6951 |
| 70 | 330 | 0.0009870 | 227.5 | 296.6 | 0.7570 |
| 70 | 335 | 0.0009895 | 247.6 | 316.9 | 0.8179 |
| 70 | 340 | 0.0009921 | 267.7 | 337.2 | 0.8780 |
| 70 | 345 | 0.0009947 | 287.8 | 357.5 | 0.9373 |
| 70 | 350 | 0.0009975 | 308.0 | 377.8 | 0.9958 |

Thermodynamic Properties of Supercritical Steam

| p | t | v | u | h | s |
|---|---|---|---|---|---|
| 70 | 355 | 0.001000 | 328.1 | 398.1 | 1.053 |
| 70 | 360 | 0.001003 | 348.3 | 418.5 | 1.110 |
| 70 | 365 | 0.001007 | 368.4 | 438.9 | 1.167 |
| 70 | 370 | 0.001010 | 388.6 | 459.3 | 1.222 |
| 70 | 375 | 0.001013 | 408.8 | 479.7 | 1.277 |
| 70 | 380 | 0.001016 | 429.0 | 500.1 | 1.331 |
| 70 | 385 | 0.001020 | 449.2 | 520.6 | 1.385 |
| 70 | 390 | 0.001024 | 469.4 | 541.1 | 1.437 |
| 70 | 395 | 0.001027 | 489.6 | 561.6 | 1.490 |
| 70 | 400 | 0.001031 | 509.9 | 582.1 | 1.541 |
| 70 | 405 | 0.001035 | 530.2 | 602.6 | 1.592 |
| 70 | 410 | 0.001039 | 550.5 | 623.2 | 1.643 |
| 70 | 415 | 0.001043 | 570.8 | 643.8 | 1.693 |
| 70 | 420 | 0.001048 | 591.1 | 664.5 | 1.742 |
| 70 | 425 | 0.001052 | 611.5 | 685.2 | 1.791 |
| 70 | 430 | 0.001057 | 631.9 | 705.9 | 1.840 |
| 70 | 435 | 0.001061 | 652.3 | 726.6 | 1.888 |
| 70 | 440 | 0.001066 | 672.8 | 747.4 | 1.935 |
| 70 | 445 | 0.001071 | 693.2 | 768.2 | 1.982 |
| 70 | 450 | 0.001076 | 713.7 | 789.1 | 2.029 |

| p | t | v | u | h | s |
|---|---|---|---|---|---|
| 70 | 455 | 0.001081 | 734.3 | 809.9 | 2.075 |
| 70 | 460 | 0.001086 | 754.8 | 830.9 | 2.121 |
| 70 | 465 | 0.001092 | 775.4 | 851.9 | 2.166 |
| 70 | 470 | 0.001097 | 796.1 | 872.9 | 2.211 |
| 70 | 475 | 0.001103 | 816.8 | 894.0 | 2.256 |
| 70 | 480 | 0.001109 | 837.5 | 915.1 | 2.300 |
| 70 | 485 | 0.001115 | 858.3 | 936.3 | 2.344 |
| 70 | 490 | 0.001121 | 879.1 | 957.6 | 2.388 |
| 70 | 495 | 0.001127 | 900.0 | 978.9 | 2.431 |
| 70 | 500 | 0.001134 | 920.9 | 1000.3 | 2.474 |
| 70 | 505 | 0.001141 | 941.9 | 1021.8 | 2.517 |
| 70 | 510 | 0.001148 | 963.0 | 1043.3 | 2.559 |
| 70 | 515 | 0.001155 | 984.1 | 1064.9 | 2.601 |
| 70 | 520 | 0.001162 | 1005.3 | 1086.6 | 2.643 |
| 70 | 525 | 0.001169 | 1026.6 | 1108.4 | 2.685 |
| 70 | 530 | 0.001177 | 1047.9 | 1130.3 | 2.726 |
| 70 | 535 | 0.001185 | 1069.3 | 1152.3 | 2.768 |
| 70 | 540 | 0.001193 | 1090.8 | 1174.4 | 2.809 |
| 70 | 545 | 0.001202 | 1112.4 | 1196.6 | 2.850 |
| 70 | 550 | 0.001210 | 1134.1 | 1218.8 | 2.890 |

| p | t | v | u | h | s |
|---|---|---|---|---|---|
| 70 | 555 | 0.001219 | 1155.9 | 1241.3 | 2.931 |
| 70 | 560 | 0.001228 | 1177.8 | 1263.8 | 2.971 |
| 70 | 565 | 0.001238 | 1199.8 | 1286.5 | 3.012 |
| 70 | 570 | 0.001248 | 1221.9 | 1309.2 | 3.052 |
| 70 | 575 | 0.001258 | 1244.1 | 1332.2 | 3.092 |
| 70 | 580 | 0.001268 | 1266.5 | 1355.3 | 3.132 |
| 70 | 585 | 0.001279 | 1289.0 | 1378.5 | 3.172 |
| 70 | 590 | 0.001291 | 1311.6 | 1401.9 | 3.211 |
| 70 | 595 | 0.001302 | 1334.3 | 1425.5 | 3.251 |
| 70 | 600 | 0.001315 | 1357.2 | 1449.2 | 3.291 |
| 70 | 605 | 0.001327 | 1380.2 | 1473.1 | 3.331 |
| 70 | 610 | 0.001340 | 1403.4 | 1497.2 | 3.370 |
| 70 | 615 | 0.001354 | 1426.8 | 1521.5 | 3.410 |
| 70 | 620 | 0.001368 | 1450.3 | 1546.0 | 3.450 |
| 70 | 630 | 0.001398 | 1497.8 | 1595.7 | 3.529 |
| 70 | 635 | 0.001414 | 1521.8 | 1620.8 | 3.569 |
| 70 | 640 | 0.001431 | 1546.1 | 1646.3 | 3.609 |
| 70 | 645 | 0.001449 | 1570.6 | 1672.1 | 3.649 |
| 70 | 650 | 0.001468 | 1595.4 | 1698.1 | 3.689 |
| 70 | 655 | 0.001487 | 1620.4 | 1724.5 | 3.730 |

| p | t | v | u | h | s |
|---|---|---|---|---|---|
| 70 | 660 | 0.001507 | 1645.6 | 1751.1 | 3.770 |
| 70 | 665 | 0.001529 | 1671.1 | 1778.1 | 3.811 |
| 70 | 670 | 0.001551 | 1696.9 | 1805.5 | 3.852 |
| 70 | 675 | 0.001575 | 1722.9 | 1833.2 | 3.893 |
| 70 | 680 | 0.001601 | 1749.3 | 1861.3 | 3.935 |
| 70 | 685 | 0.001627 | 1775.9 | 1889.9 | 3.976 |
| 70 | 690 | 0.001655 | 1803.0 | 1918.8 | 4.019 |
| 70 | 695 | 0.001685 | 1830.3 | 1948.3 | 4.061 |
| 70 | 700 | 0.001717 | 1858.0 | 1978.2 | 4.104 |
| 70 | 705 | 0.001751 | 1886.1 | 2008.6 | 4.147 |
| 70 | 710 | 0.001787 | 1914.5 | 2039.6 | 4.191 |
| 70 | 715 | 0.001825 | 1943.3 | 2071.0 | 4.235 |
| 70 | 720 | 0.001865 | 1972.4 | 2103.0 | 4.280 |
| 70 | 725 | 0.001908 | 2001.9 | 2135.5 | 4.325 |
| 70 | 730 | 0.001954 | 2031.7 | 2168.5 | 4.370 |
| 70 | 735 | 0.002002 | 2061.8 | 2201.9 | 4.416 |
| 70 | 740 | 0.002053 | 2092.0 | 2235.8 | 4.462 |
| 70 | 745 | 0.002108 | 2122.5 | 2270.0 | 4.508 |
| 70 | 750 | 0.002164 | 2153.0 | 2304.5 | 4.554 |
| 70 | 755 | 0.002224 | 2183.6 | 2339.3 | 4.600 |

Thermodynamic Properties of Supercritical Steam

| p | t | v | u | h | s |
|---|---|---|---|---|---|
| 70 | 760 | 0.002287 | 2214.2 | 2374.3 | 4.646 |
| 70 | 765 | 0.002352 | 2244.6 | 2409.3 | 4.692 |
| 70 | 770 | 0.002419 | 2274.9 | 2444.2 | 4.738 |
| 70 | 775 | 0.002489 | 2304.9 | 2479.1 | 4.783 |
| 70 | 780 | 0.002561 | 2334.5 | 2513.8 | 4.827 |
| 70 | 785 | 0.002634 | 2363.8 | 2548.1 | 4.871 |
| 70 | 790 | 0.002709 | 2392.5 | 2582.2 | 4.915 |
| 70 | 795 | 0.002785 | 2420.8 | 2615.8 | 4.957 |
| 70 | 800 | 0.002862 | 2448.5 | 2648.9 | 4.998 |
| 70 | 805 | 0.002940 | 2475.7 | 2681.5 | 5.039 |
| 70 | 810 | 0.003017 | 2502.3 | 2713.5 | 5.079 |
| 70 | 815 | 0.003096 | 2528.3 | 2745.0 | 5.117 |
| 70 | 820 | 0.003174 | 2553.7 | 2775.9 | 5.155 |
| 70 | 825 | 0.003252 | 2578.5 | 2806.1 | 5.192 |
| 70 | 830 | 0.003330 | 2602.7 | 2835.8 | 5.228 |
| 70 | 835 | 0.003407 | 2626.3 | 2864.7 | 5.263 |
| 70 | 840 | 0.003484 | 2649.3 | 2893.1 | 5.297 |
| 70 | 845 | 0.003560 | 2671.8 | 2920.9 | 5.330 |
| 70 | 850 | 0.003635 | 2693.7 | 2948.2 | 5.362 |
| 70 | 855 | 0.003710 | 2715.2 | 2974.9 | 5.393 |

| p | t | v | u | h | s |
|---|---|---|---|---|---|
| 70 | 860 | 0.003784 | 2736.2 | 3001.1 | 5.424 |
| 70 | 865 | 0.003857 | 2756.7 | 3026.7 | 5.453 |
| 70 | 870 | 0.003930 | 2776.8 | 3051.9 | 5.482 |
| 70 | 875 | 0.004001 | 2796.5 | 3076.6 | 5.511 |
| 70 | 880 | 0.004072 | 2815.8 | 3100.8 | 5.538 |
| 70 | 885 | 0.004142 | 2834.6 | 3124.6 | 5.565 |
| 70 | 890 | 0.004211 | 2853.2 | 3147.9 | 5.591 |
| 70 | 895 | 0.004279 | 2871.4 | 3170.9 | 5.617 |
| 70 | 900 | 0.004346 | 2889.2 | 3193.5 | 5.642 |
| 70 | 905 | 0.004413 | 2906.8 | 3215.7 | 5.667 |
| 70 | 910 | 0.004479 | 2924.0 | 3237.5 | 5.691 |
| 70 | 915 | 0.004544 | 2941.0 | 3259.1 | 5.715 |
| 70 | 920 | 0.004608 | 2957.8 | 3280.3 | 5.738 |
| 70 | 925 | 0.004672 | 2974.3 | 3301.3 | 5.761 |
| 70 | 930 | 0.004735 | 2990.5 | 3322.0 | 5.783 |
| 70 | 935 | 0.004797 | 3006.6 | 3342.4 | 5.805 |
| 70 | 940 | 0.004859 | 3022.4 | 3362.5 | 5.826 |
| 70 | 945 | 0.004920 | 3038.1 | 3382.4 | 5.847 |
| 70 | 950 | 0.004980 | 3053.5 | 3402.1 | 5.868 |
| 70 | 955 | 0.005040 | 3068.8 | 3421.6 | 5.889 |

Thermodynamic Properties of Supercritical Steam

| p | t | v | u | h | s |
|---|---|---|---|---|---|
| 70 | 960 | 0.005099 | 3083.9 | 3440.8 | 5.909 |
| 70 | 965 | 0.005158 | 3098.8 | 3459.8 | 5.928 |
| 70 | 970 | 0.005216 | 3113.6 | 3478.7 | 5.948 |
| 70 | 975 | 0.005273 | 3128.2 | 3497.3 | 5.967 |
| 70 | 980 | 0.005330 | 3142.7 | 3515.8 | 5.986 |
| 70 | 985 | 0.005387 | 3157.0 | 3534.1 | 6.005 |
| 70 | 990 | 0.005443 | 3171.3 | 3552.3 | 6.023 |
| 70 | 995 | 0.005498 | 3185.4 | 3570.3 | 6.041 |
| 70 | 1000 | 0.005554 | 3199.3 | 3588.1 | 6.059 |
| 70 | 1005 | 0.005608 | 3213.2 | 3605.8 | 6.077 |
| 70 | 1010 | 0.005663 | 3227.0 | 3623.4 | 6.094 |
| 70 | 1015 | 0.005717 | 3240.6 | 3640.8 | 6.111 |
| 70 | 1020 | 0.005770 | 3254.2 | 3658.1 | 6.128 |
| 70 | 1025 | 0.005823 | 3267.7 | 3675.3 | 6.145 |
| 70 | 1030 | 0.005876 | 3281.1 | 3692.4 | 6.162 |
| 70 | 1035 | 0.005928 | 3294.4 | 3709.4 | 6.178 |
| 70 | 1040 | 0.005980 | 3307.6 | 3726.2 | 6.194 |
| 70 | 1045 | 0.006032 | 3320.8 | 3743.0 | 6.211 |
| 70 | 1050 | 0.006083 | 3333.9 | 3759.7 | 6.226 |
| 70 | 1055 | 0.006134 | 3346.9 | 3776.3 | 6.242 |

| p | t | v | u | h | s |
|---|---|---|---|---|---|
| 70 | 1060 | 0.006185 | 3359.8 | 3792.8 | 6.258 |
| 70 | 1065 | 0.006235 | 3372.7 | 3809.2 | 6.273 |
| 70 | 1070 | 0.006285 | 3385.6 | 3825.6 | 6.289 |

Pressure, p = 75 Mpa

| | | | | | |
|---|---|---|---|---|---|
| 75 | 275 | 0.0009666 | 7.398 | 79.90 | 0.02261 |
| 75 | 280 | 0.0009676 | 27.19 | 99.75 | 0.09417 |
| 75 | 285 | 0.0009687 | 47.03 | 119.7 | 0.1647 |
| 75 | 290 | 0.0009700 | 66.92 | 139.7 | 0.2343 |
| 75 | 295 | 0.0009714 | 86.85 | 159.7 | 0.3028 |
| 75 | 300 | 0.0009730 | 106.8 | 179.8 | 0.3702 |
| 75 | 305 | 0.0009747 | 126.8 | 199.9 | 0.4367 |
| 75 | 310 | 0.0009766 | 146.8 | 220.0 | 0.5022 |
| 75 | 315 | 0.0009786 | 166.8 | 240.2 | 0.5667 |
| 75 | 320 | 0.0009807 | 186.8 | 260.4 | 0.6302 |
| 75 | 325 | 0.0009829 | 206.8 | 280.5 | 0.6928 |
| 75 | 330 | 0.0009852 | 226.9 | 300.8 | 0.7546 |
| 75 | 335 | 0.0009877 | 246.9 | 321.0 | 0.8154 |
| 75 | 340 | 0.0009902 | 267.0 | 341.2 | 0.8754 |

Thermodynamic Properties of Supercritical Steam

| p | t | v | u | h | s |
|---|---|---|---|---|---|
| 75 | 345 | 0.0009929 | 287.0 | 361.5 | 0.9346 |
| 75 | 350 | 0.0009956 | 307.1 | 381.8 | 0.9929 |
| 75 | 355 | 0.0009985 | 327.2 | 402.1 | 1.051 |
| 75 | 360 | 0.001001 | 347.3 | 422.4 | 1.107 |
| 75 | 365 | 0.001005 | 367.4 | 442.7 | 1.163 |
| 75 | 370 | 0.001008 | 387.5 | 463.1 | 1.219 |
| 75 | 375 | 0.001011 | 407.6 | 483.5 | 1.274 |
| 75 | 380 | 0.001014 | 427.8 | 503.9 | 1.328 |
| 75 | 385 | 0.001018 | 447.9 | 524.3 | 1.381 |
| 75 | 390 | 0.001022 | 468.1 | 544.7 | 1.434 |
| 75 | 395 | 0.001025 | 488.3 | 565.2 | 1.486 |
| 75 | 400 | 0.001029 | 508.5 | 585.7 | 1.537 |
| 75 | 405 | 0.001033 | 528.7 | 606.2 | 1.588 |
| 75 | 410 | 0.001037 | 549.0 | 626.7 | 1.639 |
| 75 | 415 | 0.001041 | 569.2 | 647.3 | 1.689 |
| 75 | 420 | 0.001045 | 589.5 | 667.9 | 1.738 |
| 75 | 425 | 0.001050 | 609.8 | 688.5 | 1.787 |
| 75 | 430 | 0.001054 | 630.1 | 709.2 | 1.835 |
| 75 | 435 | 0.001059 | 650.5 | 729.9 | 1.883 |
| 75 | 440 | 0.001063 | 670.8 | 750.6 | 1.930 |

| p | t | v | u | h | s |
|---|---|---|---|---|---|
| 75 | 445 | 0.001068 | 691.2 | 771.4 | 1.977 |
| 75 | 450 | 0.001073 | 711.7 | 792.2 | 2.024 |
| 75 | 455 | 0.001078 | 732.1 | 813.0 | 2.070 |
| 75 | 460 | 0.001083 | 752.6 | 833.9 | 2.115 |
| 75 | 465 | 0.001089 | 773.1 | 854.8 | 2.161 |
| 75 | 470 | 0.001094 | 793.7 | 875.8 | 2.206 |
| 75 | 475 | 0.001100 | 814.3 | 896.8 | 2.250 |
| 75 | 480 | 0.001106 | 834.9 | 917.9 | 2.294 |
| 75 | 485 | 0.001111 | 855.6 | 939.0 | 2.338 |
| 75 | 490 | 0.001118 | 876.4 | 960.2 | 2.381 |
| 75 | 495 | 0.001124 | 897.1 | 981.4 | 2.424 |
| 75 | 500 | 0.001130 | 918.0 | 1002.7 | 2.467 |
| 75 | 505 | 0.001137 | 938.9 | 1024.1 | 2.510 |
| 75 | 510 | 0.001143 | 959.8 | 1045.5 | 2.552 |
| 75 | 515 | 0.001150 | 980.8 | 1067.1 | 2.594 |
| 75 | 520 | 0.001157 | 1001.9 | 1088.7 | 2.636 |
| 75 | 525 | 0.001165 | 1023.0 | 1110.3 | 2.677 |
| 75 | 530 | 0.001172 | 1044.2 | 1132.1 | 2.719 |
| 75 | 535 | 0.001180 | 1065.5 | 1154.0 | 2.760 |
| 75 | 540 | 0.001188 | 1086.8 | 1175.9 | 2.800 |

Thermodynamic Properties of Supercritical Steam

| p | t | v | u | h | s |
|---|---|---|---|---|---|
| 75 | 545 | 0.001196 | 1108.3 | 1198.0 | 2.841 |
| 75 | 550 | 0.001204 | 1129.8 | 1220.1 | 2.882 |
| 75 | 555 | 0.001213 | 1151.4 | 1242.4 | 2.922 |
| 75 | 560 | 0.001222 | 1173.1 | 1264.8 | 2.962 |
| 75 | 565 | 0.001231 | 1194.9 | 1287.2 | 3.002 |
| 75 | 570 | 0.001241 | 1216.8 | 1309.9 | 3.042 |
| 75 | 575 | 0.001250 | 1238.8 | 1332.6 | 3.082 |
| 75 | 580 | 0.001261 | 1261.0 | 1355.5 | 3.121 |
| 75 | 585 | 0.001271 | 1283.2 | 1378.5 | 3.161 |
| 75 | 590 | 0.001282 | 1305.6 | 1401.7 | 3.200 |
| 75 | 595 | 0.001293 | 1328.0 | 1425.0 | 3.240 |
| 75 | 600 | 0.001305 | 1350.6 | 1448.5 | 3.279 |
| 75 | 605 | 0.001317 | 1373.4 | 1472.1 | 3.318 |
| 75 | 610 | 0.001329 | 1396.3 | 1496.0 | 3.357 |
| 75 | 615 | 0.001342 | 1419.3 | 1519.9 | 3.396 |
| 75 | 620 | 0.001356 | 1442.4 | 1544.1 | 3.436 |
| 75 | 630 | 0.001384 | 1489.1 | 1593.0 | 3.514 |
| 75 | 635 | 0.001399 | 1512.8 | 1617.7 | 3.553 |
| 75 | 640 | 0.001415 | 1536.6 | 1642.7 | 3.592 |
| 75 | 645 | 0.001432 | 1560.6 | 1668.0 | 3.631 |

| p | t | v | u | h | s |
|---|---|---|---|---|---|
| 75 | 650 | 0.001449 | 1584.9 | 1693.5 | 3.671 |
| 75 | 655 | 0.001467 | 1609.3 | 1719.3 | 3.710 |
| 75 | 660 | 0.001486 | 1633.9 | 1745.3 | 3.750 |
| 75 | 665 | 0.001506 | 1658.8 | 1771.7 | 3.790 |
| 75 | 670 | 0.001526 | 1683.8 | 1798.3 | 3.830 |
| 75 | 675 | 0.001548 | 1709.1 | 1825.2 | 3.870 |
| 75 | 680 | 0.001571 | 1734.7 | 1852.5 | 3.910 |
| 75 | 685 | 0.001595 | 1760.5 | 1880.1 | 3.950 |
| 75 | 690 | 0.001620 | 1786.6 | 1908.1 | 3.991 |
| 75 | 695 | 0.001647 | 1812.9 | 1936.4 | 4.032 |
| 75 | 700 | 0.001675 | 1839.6 | 1965.2 | 4.073 |
| 75 | 705 | 0.001705 | 1866.5 | 1994.4 | 4.115 |
| 75 | 710 | 0.001736 | 1893.7 | 2023.9 | 4.157 |
| 75 | 715 | 0.001769 | 1921.3 | 2053.9 | 4.199 |
| 75 | 720 | 0.001804 | 1949.0 | 2084.3 | 4.241 |
| 75 | 725 | 0.001841 | 1977.1 | 2115.2 | 4.284 |
| 75 | 730 | 0.001880 | 2005.4 | 2146.4 | 4.327 |
| 75 | 735 | 0.001921 | 2034.0 | 2178.1 | 4.370 |
| 75 | 740 | 0.001964 | 2062.8 | 2210.1 | 4.413 |
| 75 | 745 | 0.002010 | 2091.7 | 2242.4 | 4.457 |

| p | t | v | u | h | s |
|---|---|---|---|---|---|
| 75 | 750 | 0.002058 | 2120.7 | 2275.0 | 4.500 |
| 75 | 755 | 0.002108 | 2149.9 | 2307.9 | 4.544 |
| 75 | 760 | 0.002160 | 2179.0 | 2341.0 | 4.588 |
| 75 | 765 | 0.002215 | 2208.1 | 2374.2 | 4.631 |
| 75 | 770 | 0.002271 | 2237.2 | 2407.5 | 4.675 |
| 75 | 775 | 0.002330 | 2266.0 | 2440.8 | 4.718 |
| 75 | 780 | 0.002391 | 2294.7 | 2474.0 | 4.761 |
| 75 | 785 | 0.002453 | 2323.2 | 2507.1 | 4.803 |
| 75 | 790 | 0.002517 | 2351.3 | 2540.0 | 4.845 |
| 75 | 795 | 0.002582 | 2379.0 | 2572.7 | 4.886 |
| 75 | 800 | 0.002648 | 2406.4 | 2605.0 | 4.926 |
| 75 | 805 | 0.002716 | 2433.4 | 2637.0 | 4.966 |
| 75 | 810 | 0.002784 | 2459.8 | 2668.6 | 5.005 |
| 75 | 815 | 0.002853 | 2485.9 | 2699.8 | 5.044 |
| 75 | 820 | 0.002922 | 2511.3 | 2730.5 | 5.081 |
| 75 | 825 | 0.002992 | 2536.4 | 2760.8 | 5.118 |
| 75 | 830 | 0.003062 | 2561.0 | 2790.6 | 5.154 |
| 75 | 835 | 0.003131 | 2585.0 | 2819.9 | 5.189 |
| 75 | 840 | 0.003201 | 2608.5 | 2848.6 | 5.224 |
| 75 | 845 | 0.003270 | 2631.5 | 2876.8 | 5.257 |

| p | t | v | u | h | s |
|---|---|---|---|---|---|
| 75 | 850 | 0.003339 | 2654.0 | 2904.4 | 5.290 |
| 75 | 855 | 0.003408 | 2676.1 | 2931.6 | 5.322 |
| 75 | 860 | 0.003476 | 2697.7 | 2958.3 | 5.353 |
| 75 | 865 | 0.003543 | 2718.8 | 2984.6 | 5.383 |
| 75 | 870 | 0.003610 | 2739.6 | 3010.4 | 5.413 |
| 75 | 875 | 0.003677 | 2759.9 | 3035.7 | 5.442 |
| 75 | 880 | 0.003743 | 2779.8 | 3060.5 | 5.470 |
| 75 | 885 | 0.003808 | 2799.4 | 3085.0 | 5.498 |
| 75 | 890 | 0.003873 | 2818.5 | 3109.0 | 5.525 |
| 75 | 895 | 0.003937 | 2837.3 | 3132.6 | 5.552 |
| 75 | 900 | 0.004000 | 2855.8 | 3155.8 | 5.577 |
| 75 | 905 | 0.004063 | 2874.0 | 3178.7 | 5.603 |
| 75 | 910 | 0.004125 | 2891.9 | 3201.2 | 5.628 |
| 75 | 915 | 0.004186 | 2909.4 | 3223.4 | 5.652 |
| 75 | 920 | 0.004247 | 2926.7 | 3245.2 | 5.676 |
| 75 | 925 | 0.004307 | 2943.8 | 3266.8 | 5.699 |
| 75 | 930 | 0.004366 | 2960.6 | 3288.1 | 5.722 |
| 75 | 935 | 0.004425 | 2977.2 | 3309.1 | 5.744 |
| 75 | 940 | 0.004483 | 2993.5 | 3329.8 | 5.767 |
| 75 | 945 | 0.004541 | 3009.7 | 3350.3 | 5.788 |

Thermodynamic Properties of Supercritical Steam

| p | t | v | u | h | s |
|---|---|---|---|---|---|
| 75 | 950 | 0.004598 | 3025.6 | 3370.5 | 5.810 |
| 75 | 955 | 0.004655 | 3041.3 | 3390.5 | 5.831 |
| 75 | 960 | 0.004711 | 3056.9 | 3410.2 | 5.851 |
| 75 | 965 | 0.004767 | 3072.3 | 3429.8 | 5.872 |
| 75 | 970 | 0.004822 | 3087.5 | 3449.1 | 5.892 |
| 75 | 975 | 0.004876 | 3102.5 | 3468.3 | 5.911 |
| 75 | 980 | 0.004930 | 3117.4 | 3487.2 | 5.931 |
| 75 | 985 | 0.004984 | 3132.2 | 3506.0 | 5.950 |
| 75 | 990 | 0.005037 | 3146.8 | 3524.6 | 5.969 |
| 75 | 995 | 0.005090 | 3161.3 | 3543.0 | 5.987 |
| 75 | 1000 | 0.005142 | 3175.6 | 3561.3 | 6.005 |
| 75 | 1005 | 0.005194 | 3189.8 | 3579.4 | 6.024 |
| 75 | 1010 | 0.005246 | 3203.9 | 3597.4 | 6.041 |
| 75 | 1015 | 0.005297 | 3217.9 | 3615.2 | 6.059 |
| 75 | 1020 | 0.005348 | 3231.8 | 3632.9 | 6.076 |
| 75 | 1025 | 0.005398 | 3245.6 | 3650.5 | 6.094 |
| 75 | 1030 | 0.005448 | 3259.3 | 3667.9 | 6.111 |
| 75 | 1035 | 0.005498 | 3272.9 | 3685.3 | 6.127 |
| 75 | 1040 | 0.005547 | 3286.5 | 3702.5 | 6.144 |
| 75 | 1045 | 0.005596 | 3299.9 | 3719.6 | 6.160 |

| p | t | v | u | h | s |
|---|---|---|---|---|---|
| 75 | 1050 | 0.005645 | 3313.3 | 3736.7 | 6.177 |
| 75 | 1055 | 0.005693 | 3326.6 | 3753.6 | 6.193 |
| 75 | 1060 | 0.005741 | 3339.8 | 3770.4 | 6.209 |
| 75 | 1065 | 0.005789 | 3353.0 | 3787.2 | 6.224 |
| 75 | 1070 | 0.005837 | 3366.1 | 3803.9 | 6.240 |

Pressure, p = 80 Mpa

| | | | | | |
|---|---|---|---|---|---|
| 80 | 275 | 0.0009647 | 7.312 | 84.49 | 0.02175 |
| 80 | 280 | 0.0009656 | 27.04 | 104.3 | 0.09312 |
| 80 | 285 | 0.0009668 | 46.83 | 124.2 | 0.1635 |
| 80 | 290 | 0.0009681 | 66.67 | 144.1 | 0.2329 |
| 80 | 295 | 0.0009696 | 86.54 | 164.1 | 0.3012 |
| 80 | 300 | 0.0009712 | 106.4 | 184.1 | 0.3686 |
| 80 | 305 | 0.0009729 | 126.4 | 204.2 | 0.4349 |
| 80 | 310 | 0.0009748 | 146.3 | 224.3 | 0.5002 |
| 80 | 315 | 0.0009768 | 166.3 | 244.4 | 0.5646 |
| 80 | 320 | 0.0009789 | 186.3 | 264.6 | 0.6281 |
| 80 | 325 | 0.0009811 | 206.2 | 284.7 | 0.6906 |
| 80 | 330 | 0.0009834 | 226.2 | 304.9 | 0.7522 |
| 80 | 335 | 0.0009858 | 246.2 | 325.1 | 0.8129 |

| p | t | v | u | h | s |
|---|---|---|---|---|---|
| 80 | 340 | 0.0009884 | 266.2 | 345.3 | 0.8728 |
| 80 | 345 | 0.0009910 | 286.2 | 365.5 | 0.9319 |
| 80 | 350 | 0.0009938 | 306.3 | 385.8 | 0.9901 |
| 80 | 355 | 0.0009966 | 326.3 | 406.0 | 1.048 |
| 80 | 360 | 0.001000 | 346.4 | 426.3 | 1.104 |
| 80 | 365 | 0.001003 | 366.4 | 446.6 | 1.160 |
| 80 | 370 | 0.001006 | 386.5 | 466.9 | 1.216 |
| 80 | 375 | 0.001009 | 406.6 | 487.3 | 1.270 |
| 80 | 380 | 0.001012 | 426.6 | 507.6 | 1.324 |
| 80 | 385 | 0.001016 | 446.7 | 528.0 | 1.377 |
| 80 | 390 | 0.001019 | 466.9 | 548.4 | 1.430 |
| 80 | 395 | 0.001023 | 487.0 | 568.8 | 1.482 |
| 80 | 400 | 0.001027 | 507.1 | 589.3 | 1.534 |
| 80 | 405 | 0.001031 | 527.3 | 609.8 | 1.584 |
| 80 | 410 | 0.001035 | 547.5 | 630.3 | 1.635 |
| 80 | 415 | 0.001039 | 567.7 | 650.8 | 1.684 |
| 80 | 420 | 0.001043 | 587.9 | 671.3 | 1.734 |
| 80 | 425 | 0.001047 | 608.1 | 691.9 | 1.782 |
| 80 | 430 | 0.001052 | 628.4 | 712.5 | 1.831 |
| 80 | 435 | 0.001056 | 648.7 | 733.2 | 1.878 |

| p | t | v | u | h | s |
|---|---|---|---|---|---|
| 80 | 440 | 0.001061 | 669.0 | 753.8 | 1.926 |
| 80 | 445 | 0.001066 | 689.3 | 774.5 | 1.972 |
| 80 | 450 | 0.001070 | 709.6 | 795.3 | 2.019 |
| 80 | 455 | 0.001075 | 730.0 | 816.1 | 2.065 |
| 80 | 460 | 0.001080 | 750.4 | 836.9 | 2.110 |
| 80 | 465 | 0.001086 | 770.9 | 857.7 | 2.155 |
| 80 | 470 | 0.001091 | 791.4 | 878.7 | 2.200 |
| 80 | 475 | 0.001097 | 811.9 | 899.6 | 2.244 |
| 80 | 480 | 0.001102 | 832.4 | 920.6 | 2.288 |
| 80 | 485 | 0.001108 | 853.0 | 941.7 | 2.332 |
| 80 | 490 | 0.001114 | 873.7 | 962.8 | 2.375 |
| 80 | 495 | 0.001120 | 894.3 | 983.9 | 2.418 |
| 80 | 500 | 0.001126 | 915.1 | 1005.2 | 2.461 |
| 80 | 505 | 0.001133 | 935.8 | 1026.5 | 2.503 |
| 80 | 510 | 0.001139 | 956.7 | 1047.8 | 2.545 |
| 80 | 515 | 0.001146 | 977.6 | 1069.2 | 2.587 |
| 80 | 520 | 0.001153 | 998.5 | 1090.7 | 2.629 |
| 80 | 525 | 0.001160 | 1019.5 | 1112.3 | 2.670 |
| 80 | 530 | 0.001167 | 1040.6 | 1134.0 | 2.711 |
| 80 | 535 | 0.001175 | 1061.7 | 1155.7 | 2.752 |

Thermodynamic Properties of Supercritical Steam

| p | t | v | u | h | s |
|---|---|---|---|---|---|
| 80 | 540 | 0.001183 | 1082.9 | 1177.6 | 2.793 |
| 80 | 545 | 0.001191 | 1104.2 | 1199.5 | 2.833 |
| 80 | 550 | 0.001199 | 1125.6 | 1221.5 | 2.873 |
| 80 | 555 | 0.001207 | 1147.1 | 1243.6 | 2.913 |
| 80 | 560 | 0.001216 | 1168.6 | 1265.8 | 2.953 |
| 80 | 565 | 0.001225 | 1190.2 | 1288.2 | 2.993 |
| 80 | 570 | 0.001234 | 1211.9 | 1310.6 | 3.032 |
| 80 | 575 | 0.001243 | 1233.8 | 1333.2 | 3.072 |
| 80 | 580 | 0.001253 | 1255.7 | 1355.9 | 3.111 |
| 80 | 585 | 0.001263 | 1277.7 | 1378.7 | 3.150 |
| 80 | 590 | 0.001274 | 1299.8 | 1401.7 | 3.189 |
| 80 | 595 | 0.001284 | 1322.1 | 1424.8 | 3.228 |
| 80 | 600 | 0.001296 | 1344.4 | 1448.1 | 3.267 |
| 80 | 605 | 0.001307 | 1366.9 | 1471.5 | 3.306 |
| 80 | 610 | 0.001319 | 1389.5 | 1495.0 | 3.345 |
| 80 | 615 | 0.001331 | 1412.2 | 1518.7 | 3.384 |
| 80 | 620 | 0.001344 | 1435.0 | 1542.6 | 3.422 |
| 80 | 625 | 0.001049 | 964.1 | 1048.0 | 2.589 |
| 80 | 624 | 0.001051 | 965.5 | 1049.6 | 2.591 |
| 80 | 635 | 0.001386 | 1504.3 | 1615.2 | 3.538 |

| p | t | v | u | h | s |
|---|---|---|---|---|---|
| 80 | 645 | 0.001416 | 1551.3 | 1664.6 | 3.615 |
| 80 | 650 | 0.001432 | 1575.1 | 1689.7 | 3.654 |
| 80 | 655 | 0.001449 | 1599.0 | 1714.9 | 3.693 |
| 80 | 660 | 0.001466 | 1623.2 | 1740.5 | 3.731 |
| 80 | 665 | 0.001485 | 1647.4 | 1766.2 | 3.770 |
| 80 | 670 | 0.001504 | 1671.9 | 1792.2 | 3.809 |
| 80 | 675 | 0.001524 | 1696.6 | 1818.5 | 3.848 |
| 80 | 680 | 0.001545 | 1721.5 | 1845.1 | 3.888 |
| 80 | 685 | 0.001567 | 1746.6 | 1871.9 | 3.927 |
| 80 | 690 | 0.001590 | 1771.9 | 1899.1 | 3.966 |
| 80 | 695 | 0.001614 | 1797.4 | 1926.5 | 4.006 |
| 80 | 700 | 0.001639 | 1823.2 | 1954.3 | 4.046 |
| 80 | 705 | 0.001666 | 1849.2 | 1982.5 | 4.086 |
| 80 | 710 | 0.001694 | 1875.5 | 2011.0 | 4.126 |
| 80 | 715 | 0.001723 | 1902.0 | 2039.8 | 4.167 |
| 80 | 720 | 0.001754 | 1928.7 | 2069.0 | 4.207 |
| 80 | 725 | 0.001786 | 1955.7 | 2098.5 | 4.248 |
| 80 | 730 | 0.001820 | 1982.8 | 2128.4 | 4.289 |
| 80 | 735 | 0.001856 | 2010.2 | 2158.7 | 4.331 |
| 80 | 740 | 0.001893 | 2037.7 | 2189.2 | 4.372 |

Thermodynamic Properties of Supercritical Steam

| p | t | v | u | h | s |
|---|---|---|---|---|---|
| 80 | 745 | 0.001933 | 2065.4 | 2220.0 | 4.414 |
| 80 | 750 | 0.001974 | 2093.2 | 2251.1 | 4.455 |
| 80 | 755 | 0.002017 | 2121.1 | 2282.5 | 4.497 |
| 80 | 760 | 0.002061 | 2149.1 | 2314.0 | 4.538 |
| 80 | 765 | 0.002108 | 2177.0 | 2345.7 | 4.580 |
| 80 | 770 | 0.002157 | 2205.0 | 2377.5 | 4.621 |
| 80 | 775 | 0.002207 | 2232.8 | 2409.4 | 4.663 |
| 80 | 780 | 0.002259 | 2260.5 | 2441.2 | 4.704 |
| 80 | 785 | 0.002312 | 2288.1 | 2473.1 | 4.744 |
| 80 | 790 | 0.002367 | 2315.5 | 2504.8 | 4.785 |
| 80 | 795 | 0.002424 | 2342.6 | 2536.5 | 4.825 |
| 80 | 800 | 0.002481 | 2369.4 | 2567.9 | 4.864 |
| 80 | 805 | 0.002540 | 2395.9 | 2599.1 | 4.903 |
| 80 | 810 | 0.002599 | 2422.1 | 2630.1 | 4.941 |
| 80 | 815 | 0.002660 | 2447.9 | 2660.7 | 4.979 |
| 80 | 820 | 0.002721 | 2473.3 | 2691.0 | 5.016 |
| 80 | 825 | 0.002783 | 2498.3 | 2721.0 | 5.052 |
| 80 | 830 | 0.002845 | 2522.9 | 2750.5 | 5.088 |
| 80 | 835 | 0.002908 | 2547.1 | 2779.7 | 5.123 |
| 80 | 840 | 0.002971 | 2570.9 | 2808.5 | 5.158 |

| p | t | v | u | h | s |
|---|---|---|---|---|---|
| 80 | 845 | 0.003033 | 2594.2 | 2836.9 | 5.191 |
| 80 | 850 | 0.003096 | 2617.0 | 2864.7 | 5.224 |
| 80 | 855 | 0.003159 | 2639.4 | 2892.1 | 5.256 |
| 80 | 860 | 0.003221 | 2661.4 | 2919.1 | 5.288 |
| 80 | 865 | 0.003283 | 2683.0 | 2945.7 | 5.319 |
| 80 | 870 | 0.003345 | 2704.2 | 2971.8 | 5.349 |
| 80 | 875 | 0.003406 | 2725.1 | 2997.6 | 5.378 |
| 80 | 880 | 0.003468 | 2745.5 | 3022.9 | 5.407 |
| 80 | 885 | 0.003528 | 2765.6 | 3047.8 | 5.435 |
| 80 | 890 | 0.003588 | 2785.3 | 3072.3 | 5.463 |
| 80 | 895 | 0.003648 | 2804.6 | 3096.5 | 5.490 |
| 80 | 900 | 0.003707 | 2823.6 | 3120.2 | 5.516 |
| 80 | 905 | 0.003766 | 2842.3 | 3143.6 | 5.542 |
| 80 | 910 | 0.003824 | 2860.7 | 3166.6 | 5.568 |
| 80 | 915 | 0.003882 | 2878.8 | 3189.3 | 5.593 |
| 80 | 920 | 0.003939 | 2896.6 | 3211.7 | 5.617 |
| 80 | 925 | 0.003996 | 2914.1 | 3233.8 | 5.641 |
| 80 | 930 | 0.004052 | 2931.4 | 3255.6 | 5.664 |
| 80 | 935 | 0.004108 | 2948.5 | 3277.1 | 5.687 |
| 80 | 940 | 0.004163 | 2965.3 | 3298.3 | 5.710 |

Thermodynamic Properties of Supercritical Steam

| p | t | v | u | h | s |
|---|---|---|---|---|---|
| 80 | 945 | 0.004217 | 2981.9 | 3319.3 | 5.732 |
| 80 | 950 | 0.004272 | 2998.3 | 3340.0 | 5.754 |
| 80 | 955 | 0.004325 | 3014.5 | 3360.5 | 5.776 |
| 80 | 960 | 0.004378 | 3030.5 | 3380.7 | 5.797 |
| 80 | 965 | 0.004431 | 3046.3 | 3400.8 | 5.818 |
| 80 | 970 | 0.004483 | 3061.9 | 3420.6 | 5.838 |
| 80 | 975 | 0.004535 | 3077.3 | 3440.1 | 5.858 |
| 80 | 980 | 0.004587 | 3092.6 | 3459.5 | 5.878 |
| 80 | 985 | 0.004638 | 3107.7 | 3478.7 | 5.898 |
| 80 | 990 | 0.004688 | 3122.7 | 3497.8 | 5.917 |
| 80 | 995 | 0.004738 | 3137.5 | 3516.6 | 5.936 |
| 80 | 1000 | 0.004788 | 3152.2 | 3535.3 | 5.955 |
| 80 | 1005 | 0.004837 | 3166.8 | 3553.8 | 5.973 |
| 80 | 1010 | 0.004886 | 3181.2 | 3572.1 | 5.991 |
| 80 | 1015 | 0.004935 | 3195.6 | 3590.4 | 6.009 |
| 80 | 1020 | 0.004983 | 3209.8 | 3608.4 | 6.027 |
| 80 | 1025 | 0.005031 | 3223.9 | 3626.4 | 6.045 |
| 80 | 1030 | 0.005079 | 3237.9 | 3644.2 | 6.062 |
| 80 | 1035 | 0.005126 | 3251.8 | 3661.8 | 6.079 |
| 80 | 1040 | 0.005173 | 3265.6 | 3679.4 | 6.096 |

| p | t | v | u | h | s |
|---|---|---|---|---|---|
| 80 | 1045 | 0.005219 | 3279.3 | 3696.9 | 6.113 |
| 80 | 1050 | 0.005266 | 3293.0 | 3714.2 | 6.129 |
| 80 | 1055 | 0.005312 | 3306.5 | 3731.5 | 6.146 |
| 80 | 1060 | 0.005357 | 3320.0 | 3748.6 | 6.162 |
| 80 | 1065 | 0.005403 | 3333.5 | 3765.7 | 6.178 |
| 80 | 1070 | 0.005448 | 3346.8 | 3782.7 | 6.194 |

Pressure, p = 90 Mpa

| p | t | v | u | h | s |
|---|---|---|---|---|---|
| 90 | 275 | 0.0009608 | 7.123 | 93.60 | 0.01987 |
| 90 | 280 | 0.0009619 | 26.74 | 113.3 | 0.09090 |
| 90 | 285 | 0.0009631 | 46.42 | 133.1 | 0.1610 |
| 90 | 290 | 0.0009645 | 66.16 | 153.0 | 0.2300 |
| 90 | 295 | 0.0009660 | 85.93 | 172.9 | 0.2981 |
| 90 | 300 | 0.0009676 | 105.7 | 192.8 | 0.3652 |
| 90 | 305 | 0.0009694 | 125.6 | 212.8 | 0.4313 |
| 90 | 310 | 0.0009713 | 145.4 | 232.8 | 0.4964 |
| 90 | 315 | 0.0009732 | 165.3 | 252.9 | 0.5605 |
| 90 | 320 | 0.0009753 | 185.2 | 272.9 | 0.6237 |
| 90 | 325 | 0.0009776 | 205.1 | 293.0 | 0.6860 |
| 90 | 330 | 0.0009799 | 225.0 | 313.1 | 0.7474 |

Thermodynamic Properties of Supercritical Steam

| p | t | v | u | h | s |
|---|---|---|---|---|---|
| 90 | 335 | 0.0009823 | 244.9 | 333.3 | 0.8080 |
| 90 | 340 | 0.0009848 | 264.8 | 353.4 | 0.8676 |
| 90 | 345 | 0.0009874 | 284.7 | 373.6 | 0.9265 |
| 90 | 350 | 0.0009901 | 304.6 | 393.7 | 0.9846 |
| 90 | 355 | 0.0009930 | 324.6 | 413.9 | 1.042 |
| 90 | 360 | 0.0009959 | 344.5 | 434.2 | 1.098 |
| 90 | 365 | 0.0009989 | 364.5 | 454.4 | 1.154 |
| 90 | 370 | 0.001002 | 384.5 | 474.6 | 1.209 |
| 90 | 375 | 0.001005 | 404.4 | 494.9 | 1.264 |
| 90 | 380 | 0.001008 | 424.4 | 515.2 | 1.317 |
| 90 | 385 | 0.001012 | 444.4 | 535.5 | 1.371 |
| 90 | 390 | 0.001015 | 464.4 | 555.8 | 1.423 |
| 90 | 395 | 0.001019 | 484.5 | 576.2 | 1.475 |
| 90 | 400 | 0.001023 | 504.5 | 596.5 | 1.526 |
| 90 | 405 | 0.001026 | 524.5 | 616.9 | 1.577 |
| 90 | 410 | 0.001030 | 544.6 | 637.3 | 1.627 |
| 90 | 415 | 0.001034 | 564.7 | 657.8 | 1.676 |
| 90 | 420 | 0.001038 | 584.8 | 678.2 | 1.725 |
| 90 | 425 | 0.001043 | 604.9 | 698.7 | 1.774 |
| 90 | 430 | 0.001047 | 625.0 | 719.2 | 1.822 |

| p | t | v | u | h | s |
|---|---|---|---|---|---|
| 90 | 435 | 0.001051 | 645.2 | 739.8 | 1.869 |
| 90 | 440 | 0.001056 | 665.3 | 760.3 | 1.916 |
| 90 | 445 | 0.001060 | 685.5 | 781.0 | 1.963 |
| 90 | 450 | 0.001065 | 705.7 | 801.6 | 2.009 |
| 90 | 455 | 0.001070 | 726.0 | 822.3 | 2.055 |
| 90 | 460 | 0.001075 | 746.2 | 843.0 | 2.100 |
| 90 | 465 | 0.001080 | 766.5 | 863.7 | 2.145 |
| 90 | 470 | 0.001085 | 786.9 | 884.5 | 2.189 |
| 90 | 475 | 0.001090 | 807.2 | 905.3 | 2.233 |
| 90 | 480 | 0.001096 | 827.6 | 926.2 | 2.277 |
| 90 | 485 | 0.001101 | 848.0 | 947.1 | 2.320 |
| 90 | 490 | 0.001107 | 868.5 | 968.1 | 2.364 |
| 90 | 495 | 0.001113 | 889.0 | 989.1 | 2.406 |
| 90 | 500 | 0.001119 | 909.5 | 1010.2 | 2.449 |
| 90 | 505 | 0.001125 | 930.1 | 1031.3 | 2.491 |
| 90 | 510 | 0.001131 | 950.7 | 1052.5 | 2.532 |
| 90 | 515 | 0.001138 | 971.4 | 1073.8 | 2.574 |
| 90 | 520 | 0.001145 | 992.1 | 1095.1 | 2.615 |
| 90 | 525 | 0.001151 | 1012.9 | 1116.5 | 2.656 |
| 90 | 530 | 0.001158 | 1033.7 | 1137.9 | 2.697 |

Thermodynamic Properties of Supercritical Steam

| p | t | v | u | h | s |
|---|---|---|---|---|---|
| 90 | 535 | 0.001165 | 1054.6 | 1159.5 | 2.737 |
| 90 | 540 | 0.001173 | 1075.5 | 1181.1 | 2.777 |
| 90 | 545 | 0.001180 | 1096.6 | 1202.8 | 2.817 |
| 90 | 550 | 0.001188 | 1117.6 | 1224.6 | 2.857 |
| 90 | 555 | 0.001196 | 1138.8 | 1246.4 | 2.897 |
| 90 | 560 | 0.001204 | 1160.0 | 1268.4 | 2.936 |
| 90 | 565 | 0.001213 | 1181.3 | 1290.5 | 2.975 |
| 90 | 570 | 0.001221 | 1202.7 | 1312.6 | 3.014 |
| 90 | 575 | 0.001230 | 1224.2 | 1334.9 | 3.053 |
| 90 | 580 | 0.001239 | 1245.7 | 1357.3 | 3.092 |
| 90 | 585 | 0.001249 | 1267.4 | 1379.7 | 3.131 |
| 90 | 590 | 0.001258 | 1289.1 | 1402.3 | 3.169 |
| 90 | 595 | 0.001268 | 1310.9 | 1425.0 | 3.207 |
| 90 | 600 | 0.001279 | 1332.8 | 1447.9 | 3.246 |
| 90 | 605 | 0.001289 | 1354.8 | 1470.8 | 3.284 |
| 90 | 610 | 0.001300 | 1376.9 | 1493.9 | 3.322 |
| 90 | 615 | 0.001312 | 1399.1 | 1517.1 | 3.360 |
| 90 | 620 | 0.001323 | 1421.4 | 1540.5 | 3.397 |
| 90 | 640 | 0.001374 | 1511.6 | 1635.2 | 3.548 |
| 90 | 645 | 0.001388 | 1534.4 | 1659.4 | 3.585 |

| p | t | v | u | h | s |
|---|---|---|---|---|---|
| 90 | 650 | 0.001402 | 1557.4 | 1683.7 | 3.623 |
| 90 | 655 | 0.001417 | 1580.6 | 1708.1 | 3.660 |
| 90 | 660 | 0.001433 | 1603.9 | 1732.8 | 3.698 |
| 90 | 665 | 0.001449 | 1627.3 | 1757.7 | 3.735 |
| 90 | 670 | 0.001466 | 1650.8 | 1782.7 | 3.773 |
| 90 | 675 | 0.001483 | 1674.5 | 1807.9 | 3.810 |
| 90 | 680 | 0.001501 | 1698.3 | 1833.4 | 3.848 |
| 90 | 685 | 0.001520 | 1722.3 | 1859.0 | 3.886 |
| 90 | 690 | 0.001539 | 1746.4 | 1884.9 | 3.923 |
| 90 | 695 | 0.001559 | 1770.7 | 1911.0 | 3.961 |
| 90 | 700 | 0.001581 | 1795.1 | 1937.4 | 3.999 |
| 90 | 705 | 0.001603 | 1819.7 | 1964.0 | 4.037 |
| 90 | 710 | 0.001626 | 1844.5 | 1990.8 | 4.074 |
| 90 | 715 | 0.001650 | 1869.4 | 2017.9 | 4.113 |
| 90 | 720 | 0.001675 | 1894.6 | 2045.3 | 4.151 |
| 90 | 725 | 0.001701 | 1919.8 | 2072.9 | 4.189 |
| 90 | 730 | 0.001728 | 1945.3 | 2100.8 | 4.227 |
| 90 | 735 | 0.001756 | 1970.8 | 2128.9 | 4.266 |
| 90 | 740 | 0.001786 | 1996.6 | 2157.3 | 4.304 |
| 90 | 745 | 0.001817 | 2022.4 | 2185.9 | 4.343 |

| p | t | v | u | h | s |
|---|---|---|---|---|---|
| 90 | 750 | 0.001849 | 2048.3 | 2214.7 | 4.381 |
| 90 | 755 | 0.001882 | 2074.3 | 2243.7 | 4.420 |
| 90 | 760 | 0.001917 | 2100.4 | 2272.9 | 4.458 |
| 90 | 765 | 0.001953 | 2126.5 | 2302.3 | 4.497 |
| 90 | 770 | 0.001990 | 2152.6 | 2331.7 | 4.535 |
| 90 | 775 | 0.002029 | 2178.7 | 2361.3 | 4.573 |
| 90 | 780 | 0.002068 | 2204.8 | 2390.9 | 4.612 |
| 90 | 785 | 0.002109 | 2230.8 | 2420.6 | 4.649 |
| 90 | 790 | 0.002152 | 2256.7 | 2450.3 | 4.687 |
| 90 | 795 | 0.002195 | 2282.4 | 2480.0 | 4.725 |
| 90 | 800 | 0.002239 | 2308.1 | 2509.6 | 4.762 |
| 90 | 805 | 0.002285 | 2333.5 | 2539.1 | 4.799 |
| 90 | 810 | 0.002331 | 2358.7 | 2568.6 | 4.835 |
| 90 | 815 | 0.002379 | 2383.8 | 2597.8 | 4.871 |
| 90 | 820 | 0.002427 | 2408.5 | 2626.9 | 4.907 |
| 90 | 825 | 0.002476 | 2433.0 | 2655.9 | 4.942 |
| 90 | 830 | 0.002525 | 2457.3 | 2684.6 | 4.976 |
| 90 | 835 | 0.002576 | 2481.2 | 2713.0 | 5.011 |
| 90 | 840 | 0.002626 | 2504.9 | 2741.2 | 5.044 |
| 90 | 845 | 0.002677 | 2528.2 | 2769.2 | 5.077 |

| p | t | v | u | h | s |
|---|---|---|---|---|---|
| 90 | 850 | 0.002729 | 2551.2 | 2796.8 | 5.110 |
| 90 | 855 | 0.002781 | 2573.9 | 2824.2 | 5.142 |
| 90 | 860 | 0.002832 | 2596.3 | 2851.2 | 5.174 |
| 90 | 865 | 0.002885 | 2618.3 | 2877.9 | 5.205 |
| 90 | 870 | 0.002937 | 2640.0 | 2904.3 | 5.235 |
| 90 | 875 | 0.002989 | 2661.4 | 2930.4 | 5.265 |
| 90 | 880 | 0.003041 | 2682.4 | 2956.1 | 5.294 |
| 90 | 885 | 0.003093 | 2703.2 | 2981.5 | 5.323 |
| 90 | 890 | 0.003145 | 2723.6 | 3006.6 | 5.351 |
| 90 | 895 | 0.003196 | 2743.7 | 3031.3 | 5.379 |
| 90 | 900 | 0.003248 | 2763.4 | 3055.7 | 5.406 |
| 90 | 905 | 0.003299 | 2782.9 | 3079.7 | 5.433 |
| 90 | 910 | 0.003350 | 2802.0 | 3103.5 | 5.459 |
| 90 | 915 | 0.003400 | 2820.9 | 3126.9 | 5.485 |
| 90 | 920 | 0.003451 | 2839.4 | 3150.0 | 5.510 |
| 90 | 925 | 0.003501 | 2857.8 | 3172.8 | 5.535 |
| 90 | 930 | 0.003551 | 2875.8 | 3195.4 | 5.559 |
| 90 | 935 | 0.003600 | 2893.7 | 3217.7 | 5.583 |
| 90 | 940 | 0.003649 | 2911.2 | 3239.7 | 5.606 |
| 90 | 945 | 0.003698 | 2928.6 | 3261.4 | 5.629 |

Thermodynamic Properties of Supercritical Steam

| p | t | v | u | h | s |
|---|---|---|---|---|---|
| 90 | 950 | 0.003747 | 2945.8 | 3283.0 | 5.652 |
| 90 | 955 | 0.003795 | 2962.7 | 3304.2 | 5.674 |
| 90 | 960 | 0.003842 | 2979.5 | 3325.3 | 5.696 |
| 90 | 965 | 0.003890 | 2996.0 | 3346.1 | 5.718 |
| 90 | 970 | 0.003937 | 3012.4 | 3366.7 | 5.739 |
| 90 | 975 | 0.003984 | 3028.5 | 3387.1 | 5.760 |
| 90 | 980 | 0.004030 | 3044.5 | 3407.2 | 5.781 |
| 90 | 985 | 0.004076 | 3060.4 | 3427.2 | 5.801 |
| 90 | 990 | 0.004122 | 3076.0 | 3446.9 | 5.821 |
| 90 | 995 | 0.004167 | 3091.5 | 3466.5 | 5.841 |
| 90 | 1000 | 0.004212 | 3106.8 | 3485.9 | 5.860 |
| 90 | 1005 | 0.004256 | 3122.0 | 3505.1 | 5.880 |
| 90 | 1010 | 0.004301 | 3137.1 | 3524.2 | 5.898 |
| 90 | 1015 | 0.004345 | 3152.0 | 3543.1 | 5.917 |
| 90 | 1020 | 0.004388 | 3166.8 | 3561.8 | 5.936 |
| 90 | 1025 | 0.004432 | 3181.5 | 3580.4 | 5.954 |
| 90 | 1030 | 0.004475 | 3196.1 | 3598.8 | 5.972 |
| 90 | 1035 | 0.004518 | 3210.5 | 3617.1 | 5.989 |
| 90 | 1040 | 0.004560 | 3224.9 | 3635.3 | 6.007 |
| 90 | 1045 | 0.004603 | 3239.1 | 3653.4 | 6.024 |

| p | t | v | u | h | s |
|---|---|---|---|---|---|
| 90 | 1050 | 0.004645 | 3253.3 | 3671.3 | 6.041 |
| 90 | 1055 | 0.004686 | 3267.3 | 3689.1 | 6.058 |
| 90 | 1060 | 0.004728 | 3281.3 | 3706.9 | 6.075 |
| 90 | 1065 | 0.004769 | 3295.3 | 3724.5 | 6.092 |
| 90 | 1070 | 0.004810 | 3309.1 | 3742.0 | 6.108 |

Pressure p = 100 Mpa

| | | | | | |
|---|---|---|---|---|---|
| 100 | 275 | 0.0009571 | 6.912 | 102.62 | 0.01781 |
| 100 | 280 | 0.0009582 | 26.42 | 122.2 | 0.08853 |
| 100 | 285 | 0.0009595 | 46.00 | 142.0 | 0.1583 |
| 100 | 290 | 0.0009609 | 65.64 | 161.7 | 0.2271 |
| 100 | 295 | 0.0009625 | 85.33 | 181.6 | 0.2949 |
| 100 | 300 | 0.0009641 | 105.0 | 201.5 | 0.3618 |
| 100 | 305 | 0.0009659 | 124.8 | 221.4 | 0.4276 |
| 100 | 310 | 0.0009678 | 144.5 | 241.3 | 0.4925 |
| 100 | 315 | 0.0009698 | 164.3 | 261.3 | 0.5564 |
| 100 | 320 | 0.0009719 | 184.1 | 281.3 | 0.6194 |
| 100 | 325 | 0.0009741 | 203.9 | 301.3 | 0.6815 |
| 100 | 330 | 0.0009764 | 223.7 | 321.4 | 0.7427 |
| 100 | 335 | 0.0009788 | 243.5 | 341.4 | 0.8030 |

| p | t | v | u | h | s |
|---|---|---|---|---|---|
| 100 | 340 | 0.0009813 | 263.4 | 361.5 | 0.8625 |
| 100 | 345 | 0.0009839 | 283.2 | 381.6 | 0.9212 |
| 100 | 350 | 0.0009866 | 303.0 | 401.7 | 0.9791 |
| 100 | 355 | 0.0009894 | 322.9 | 421.8 | 1.036 |
| 100 | 360 | 0.0009923 | 342.8 | 442.0 | 1.093 |
| 100 | 365 | 0.0009952 | 362.6 | 462.1 | 1.148 |
| 100 | 370 | 0.0009983 | 382.5 | 482.3 | 1.203 |
| 100 | 375 | 0.001001 | 402.4 | 502.5 | 1.257 |
| 100 | 380 | 0.001005 | 422.3 | 522.7 | 1.311 |
| 100 | 385 | 0.001008 | 442.2 | 543.0 | 1.364 |
| 100 | 390 | 0.001011 | 462.1 | 563.2 | 1.416 |
| 100 | 395 | 0.001015 | 482.0 | 583.5 | 1.468 |
| 100 | 400 | 0.001019 | 501.9 | 603.8 | 1.519 |
| 100 | 405 | 0.001022 | 521.9 | 624.1 | 1.569 |
| 100 | 410 | 0.001026 | 541.8 | 644.4 | 1.619 |
| 100 | 415 | 0.001030 | 561.8 | 664.8 | 1.668 |
| 100 | 420 | 0.001034 | 581.8 | 685.2 | 1.717 |
| 100 | 425 | 0.001038 | 601.8 | 705.6 | 1.765 |
| 100 | 430 | 0.001042 | 621.8 | 726.0 | 1.813 |
| 100 | 435 | 0.001046 | 641.8 | 746.4 | 1.861 |

| p | t | v | u | h | s |
|---|---|---|---|---|---|
| 100 | 440 | 0.001051 | 661.8 | 766.9 | 1.907 |
| 100 | 445 | 0.001055 | 681.9 | 787.4 | 1.954 |
| 100 | 450 | 0.001060 | 702.0 | 808.0 | 2.000 |
| 100 | 455 | 0.001065 | 722.1 | 828.5 | 2.045 |
| 100 | 460 | 0.001069 | 742.2 | 849.2 | 2.090 |
| 100 | 465 | 0.001074 | 762.4 | 869.8 | 2.135 |
| 100 | 470 | 0.001079 | 782.5 | 890.5 | 2.179 |
| 100 | 475 | 0.001084 | 802.7 | 911.2 | 2.223 |
| 100 | 480 | 0.001090 | 823.0 | 931.9 | 2.266 |
| 100 | 485 | 0.001095 | 843.2 | 952.7 | 2.309 |
| 100 | 490 | 0.001101 | 863.5 | 973.6 | 2.352 |
| 100 | 495 | 0.001106 | 883.8 | 994.5 | 2.395 |
| 100 | 500 | 0.001112 | 904.2 | 1015.4 | 2.437 |
| 100 | 505 | 0.001118 | 924.6 | 1036.4 | 2.478 |
| 100 | 510 | 0.001124 | 945.0 | 1057.4 | 2.520 |
| 100 | 515 | 0.001130 | 965.5 | 1078.5 | 2.561 |
| 100 | 520 | 0.001137 | 986.0 | 1099.7 | 2.602 |
| 100 | 525 | 0.001143 | 1006.6 | 1120.9 | 2.642 |
| 100 | 530 | 0.001150 | 1027.2 | 1142.2 | 2.683 |
| 100 | 535 | 0.001157 | 1047.9 | 1163.5 | 2.723 |

Thermodynamic Properties of Supercritical Steam

| p | t | v | u | h | s |
|---|---|---|---|---|---|
| 100 | 540 | 0.001164 | 1068.6 | 1184.9 | 2.763 |
| 100 | 545 | 0.001171 | 1089.3 | 1206.4 | 2.802 |
| 100 | 550 | 0.001178 | 1110.2 | 1228.0 | 2.842 |
| 100 | 555 | 0.001186 | 1131.1 | 1249.6 | 2.881 |
| 100 | 560 | 0.001193 | 1152.0 | 1271.4 | 2.920 |
| 100 | 565 | 0.001201 | 1173.0 | 1293.2 | 2.959 |
| 100 | 570 | 0.001210 | 1194.1 | 1315.1 | 2.997 |
| 100 | 575 | 0.001218 | 1215.3 | 1337.1 | 3.036 |
| 100 | 580 | 0.001227 | 1236.5 | 1359.1 | 3.074 |
| 100 | 585 | 0.001235 | 1257.8 | 1381.3 | 3.112 |
| 100 | 590 | 0.001244 | 1279.2 | 1403.6 | 3.150 |
| 100 | 595 | 0.001254 | 1300.6 | 1426.0 | 3.188 |
| 100 | 600 | 0.001263 | 1322.1 | 1448.5 | 3.225 |
| 100 | 605 | 0.001273 | 1343.7 | 1471.0 | 3.263 |
| 100 | 610 | 0.001283 | 1365.4 | 1493.7 | 3.300 |
| 100 | 615 | 0.001294 | 1387.1 | 1516.5 | 3.337 |
| 100 | 620 | 0.001305 | 1409.0 | 1539.4 | 3.374 |
| 100 | 645 | 0.001364 | 1519.4 | 1655.8 | 3.558 |
| 100 | 650 | 0.001377 | 1541.8 | 1679.5 | 3.595 |
| 100 | 655 | 0.001391 | 1564.3 | 1703.3 | 3.632 |

| p | t | v | u | h | s |
|---|---|---|---|---|---|
| 100 | 660 | 0.001404 | 1586.9 | 1727.3 | 3.668 |
| 100 | 665 | 0.001419 | 1609.6 | 1751.5 | 3.705 |
| 100 | 670 | 0.001434 | 1632.4 | 1775.8 | 3.741 |
| 100 | 675 | 0.001449 | 1655.3 | 1800.2 | 3.777 |
| 100 | 680 | 0.001465 | 1678.3 | 1824.8 | 3.814 |
| 100 | 685 | 0.001481 | 1701.4 | 1849.6 | 3.850 |
| 100 | 690 | 0.001498 | 1724.7 | 1874.5 | 3.886 |
| 100 | 695 | 0.001516 | 1748.0 | 1899.6 | 3.922 |
| 100 | 700 | 0.001534 | 1771.5 | 1924.9 | 3.959 |
| 100 | 705 | 0.001553 | 1795.0 | 1950.3 | 3.995 |
| 100 | 710 | 0.001573 | 1818.7 | 1976.0 | 4.031 |
| 100 | 715 | 0.001593 | 1842.6 | 2001.9 | 4.067 |
| 100 | 720 | 0.001614 | 1866.5 | 2027.9 | 4.104 |
| 100 | 725 | 0.001636 | 1890.6 | 2054.2 | 4.140 |
| 100 | 730 | 0.001659 | 1914.8 | 2080.7 | 4.176 |
| 100 | 735 | 0.001683 | 1939.0 | 2107.3 | 4.213 |
| 100 | 740 | 0.001707 | 1963.4 | 2134.2 | 4.249 |
| 100 | 745 | 0.001733 | 1987.9 | 2161.2 | 4.286 |
| 100 | 750 | 0.001759 | 2012.5 | 2188.4 | 4.322 |
| 100 | 755 | 0.001786 | 2037.1 | 2215.8 | 4.358 |

| p | t | v | u | h | s |
|---|---|---|---|---|---|
| 100 | 760 | 0.001815 | 2061.8 | 2243.3 | 4.395 |
| 100 | 765 | 0.001844 | 2086.6 | 2270.9 | 4.431 |
| 100 | 770 | 0.001874 | 2111.3 | 2298.7 | 4.467 |
| 100 | 775 | 0.001905 | 2136.1 | 2326.6 | 4.503 |
| 100 | 780 | 0.001937 | 2160.8 | 2354.5 | 4.539 |
| 100 | 785 | 0.001970 | 2185.5 | 2382.5 | 4.575 |
| 100 | 790 | 0.002004 | 2210.2 | 2410.6 | 4.611 |
| 100 | 795 | 0.002038 | 2234.8 | 2438.6 | 4.646 |
| 100 | 800 | 0.002074 | 2259.3 | 2466.7 | 4.681 |
| 100 | 805 | 0.002111 | 2283.7 | 2494.8 | 4.716 |
| 100 | 810 | 0.002148 | 2308.0 | 2522.8 | 4.751 |
| 100 | 815 | 0.002186 | 2332.1 | 2550.7 | 4.785 |
| 100 | 820 | 0.002225 | 2356.1 | 2578.6 | 4.819 |
| 100 | 825 | 0.002265 | 2379.9 | 2606.4 | 4.853 |
| 100 | 830 | 0.002305 | 2403.5 | 2634.0 | 4.887 |
| 100 | 835 | 0.002346 | 2426.9 | 2661.5 | 4.920 |
| 100 | 840 | 0.002387 | 2450.1 | 2688.8 | 4.952 |
| 100 | 845 | 0.002429 | 2473.1 | 2716.0 | 4.984 |
| 100 | 850 | 0.002471 | 2495.8 | 2743.0 | 5.016 |
| 100 | 855 | 0.002514 | 2518.3 | 2769.8 | 5.048 |

| p | t | v | u | h | s |
|---|---|---|---|---|---|
| 100 | 860 | 0.002557 | 2540.6 | 2796.3 | 5.079 |
| 100 | 865 | 0.002601 | 2562.6 | 2822.7 | 5.109 |
| 100 | 870 | 0.002645 | 2584.3 | 2848.7 | 5.139 |
| 100 | 875 | 0.002689 | 2605.8 | 2874.6 | 5.169 |
| 100 | 880 | 0.002733 | 2627.0 | 2900.3 | 5.198 |
| 100 | 885 | 0.002777 | 2648.1 | 2925.8 | 5.227 |
| 100 | 890 | 0.002822 | 2668.8 | 2951.0 | 5.255 |
| 100 | 895 | 0.002866 | 2689.3 | 2975.9 | 5.283 |
| 100 | 900 | 0.002910 | 2709.5 | 3000.5 | 5.311 |
| 100 | 905 | 0.002954 | 2729.4 | 3024.8 | 5.338 |
| 100 | 910 | 0.002999 | 2749.0 | 3048.8 | 5.364 |
| 100 | 915 | 0.003043 | 2768.3 | 3072.5 | 5.390 |
| 100 | 920 | 0.003087 | 2787.3 | 3096.0 | 5.416 |
| 100 | 925 | 0.003131 | 2806.1 | 3119.1 | 5.441 |
| 100 | 930 | 0.003175 | 2824.6 | 3142.1 | 5.466 |
| 100 | 935 | 0.003218 | 2842.9 | 3164.8 | 5.490 |
| 100 | 940 | 0.003262 | 2861.0 | 3187.2 | 5.514 |
| 100 | 945 | 0.003305 | 2878.9 | 3209.5 | 5.537 |
| 100 | 950 | 0.003349 | 2896.6 | 3231.5 | 5.561 |
| 100 | 955 | 0.003392 | 2914.1 | 3253.3 | 5.584 |

Thermodynamic Properties of Supercritical Steam

| p | t | v | u | h | s |
|---|---|---|---|---|---|
| 100 | 960 | 0.003435 | 2931.5 | 3274.9 | 5.606 |
| 100 | 965 | 0.003477 | 2948.6 | 3296.3 | 5.628 |
| 100 | 970 | 0.003520 | 2965.6 | 3317.5 | 5.650 |
| 100 | 975 | 0.003562 | 2982.3 | 3338.5 | 5.672 |
| 100 | 980 | 0.003604 | 2998.9 | 3359.3 | 5.693 |
| 100 | 985 | 0.003645 | 3015.3 | 3379.8 | 5.714 |
| 100 | 990 | 0.003686 | 3031.6 | 3400.2 | 5.735 |
| 100 | 995 | 0.003727 | 3047.7 | 3420.4 | 5.755 |
| 100 | 1000 | 0.003768 | 3063.6 | 3440.4 | 5.775 |
| 100 | 1005 | 0.003809 | 3079.4 | 3460.2 | 5.795 |
| 100 | 1010 | 0.003849 | 3095.0 | 3479.8 | 5.814 |
| 100 | 1015 | 0.003889 | 3110.4 | 3499.3 | 5.834 |
| 100 | 1020 | 0.003928 | 3125.8 | 3518.6 | 5.853 |
| 100 | 1025 | 0.003968 | 3141.0 | 3537.7 | 5.871 |
| 100 | 1030 | 0.004007 | 3156.0 | 3556.7 | 5.890 |
| 100 | 1035 | 0.004046 | 3171.0 | 3575.6 | 5.908 |
| 100 | 1040 | 0.004084 | 3185.8 | 3594.3 | 5.926 |
| 100 | 1045 | 0.004123 | 3200.5 | 3612.8 | 5.944 |
| 100 | 1050 | 0.004161 | 3215.2 | 3631.3 | 5.961 |
| 100 | 1055 | 0.004199 | 3229.7 | 3649.6 | 5.979 |

| p | t | v | u | h | s |
|-----|------|----------|--------|--------|-------|
| 100 | 1060 | 0.004237 | 3244.1 | 3667.8 | 5.996 |
| 100 | 1065 | 0.004275 | 3258.4 | 3685.9 | 6.013 |
| 100 | 1070 | 0.004312 | 3272.7 | 3703.9 | 6.030 |

www.ingramcontent.com/pod-product-compliance
Lightning Source LLC
Chambersburg PA
CBHW032022170526
45157CB00002B/809